In the Name of Sharks

FRANÇOIS SARANO

Foreword by Sandra Bessudo
Translated by Stephen Muecke
Original illustrations by Marion Sarano

T0243958

polity

Originally published in French as *Au nom des requins* by François Sarano
© Actes Sud, 2022

This English edition © Polity Press, 2024

This book is supported by the Institut français (Royaume-Uni) as part of the Burgess programme.

INSTITUT
FRANÇAIS
ROYAUME-UNI

Polity Press
65 Bridge Street
Cambridge CB2 1UR, UK

Polity Press
111 River Street
Hoboken, NJ 07030, USA

ISBN-13: 978-1-5095-5766-0 – hardback
ISBN-13: 978-1-5095-5767-7 – paperback

A catalogue record for this book is available from the British Library.

Library of Congress Control Number: 2023934605

Typeset in 11 on 14pt Warnock Pro
by Cheshire Typesetting Ltd, Cuddington, Cheshire
Printed and bound in Great Britain by T.J. Books Ltd, Padstow, Cornwall

The publisher has used its best endeavours to ensure that the URLs for external websites referred to in this book are correct and active at the time of going to press. However, the publisher has no responsibility for the websites and can make no guarantee that a site will remain live or that the content is or will remain appropriate.

Every effort has been made to trace all copyright holders, but if any have been overlooked the publisher will be pleased to include any necessary credits in any subsequent reprint or edition.

For further information on Polity, visit our website:
politybooks.com

In the Name of Sharks

Contents

Foreword by Sandra Bessudo   vi

Introduction: Giving the 'Voiceless' a Voice   1
1. A Matter of Misunderstanding: From Pliny to Disney   4
2. Shark? What Shark?   29
3. Giving Life   60
4. Inside the Shark's Head   71
5. On the Road to Personality   88
6. The Shark, Where it Belongs   104
7. The Ocean is Their Garden   134
8. Fading Silhouettes   154
9. The Confrontation   168
10. Reconciliation   197

Notes   212
List of Illustrations   249
Acknowledgements   251
Index   253

Foreword

Ⅎrançois Sarano is an oceanographer who has had the good fortune to explore the world's oceans aboard Captain Cousteau's famous *Calypso*, and he shares my passion for knowing and protecting the living creatures that inhabit our oceans. Most of us are fascinated by sharks, probably the most emblematic of them all. Some even have worshipped them as gods, since ancient times. However, their bad reputation and the fear stirred up by films such as *Jaws* have encouraged their elimination. Overfishing for their meat and fins, and accidental capture by fishing gear not intended for them, have led to the collapse of most populations. The final blow has been dealt by the insatiable demand for shark fins from an ever-growing Asian market. As recently as September 2021, Colombian customs seized several thousand fins on their way to Hong Kong. Such wasteful carnage!

Scalloped hammerhead sharks (*Sphyrna lewini*), silky sharks, thresher sharks and many others, so abundant a few years ago, are in danger of disappearing. Our scientific studies show that all are highly migratory, crossing the oceans from island to island, continent to continent, disregarding our artificial boundaries. This can be a misfortune or an opportunity! A misfortune, if these sharks, protected in one country, are massacred in the international waters they cross and in the country where they go to breed. But it can also be a great opportunity to unite countries, governments and people in a common effort to protect sharks and the oceans, for the common good of our children and their descendants. This is what we hope to achieve by working towards the creation of a protected marine corridor that would link Colombia, Ecuador, Panama and Costa Rica, all countries that are visited each year by hammerhead sharks from our Colombian Malpelo archipelago.

With his book, *In the Name of Sharks*, François Sarano, to whom I was able to offer a sample of the joys of diving in the dynamic, abundant and pristine waters of Malpelo, takes us on a whirlwind of dives, face to face with our shark friends. From these dives, sometimes surprising, always moving, he draws out his knowledge of the behaviours and relationships that sharks weave with other species. He demonstrates the importance of sharks not only to ocean ecosystems, but to us humans. François tells us how much we need to respect sharks to rebuild ourselves.

May this book make us understand the richness of our interdependencies. May this book be the advocate that raises our awareness and pushes us to protect this enchanting biodiversity. May this text be added to the testimonies of other scientists to convince us not to be the ones who cause the extinction of sharks that have survived so many geological extinctions.

Thank you, François, for this book that teaches us about sharks to make us love them, and that invites us to appreciate our good fortune in living alongside them.

Thank you, in the name of sharks!

Sandra Bessudo,
Naturalist,
former Minister of the Environment,
former president and advisor to the
Colombian vice president of the Colombian Ocean Commission,
founder of the Malpelo Foundation and other marine ecosystems,
instigator and first director of the Malpelo Wildlife Sanctuary,
declared a World Heritage Site in 2006.

Introduction:
Giving the 'Voiceless' a Voice

It was Lady Mystery who gave me the idea for this book, when she kindly granted me a remarkable *tête-à-tête* on 12 November 2006 off the coast of Mexico. With five metres of muscle and a tonne of elegance, Lady Mystery is a great white shark, *Carcharodon carcharias*, a sister to those in Steven Spielberg's *Jaws*.[1] During the shooting of the film *Oceans*,[2] we swam serenely side by side, shoulder to shoulder, eye to eye, a few centimetres apart. Two minutes of contentment and peace, an eternity of happiness! For my incredulous diving companions, as well as for Jacques Perrin and Jacques Cluzaud, the film's directors, this harmonious moment was the occasion for a radical change in the way this much-feared creature is seen. This encounter, beautifully filmed by David Reichert and Didier Noirot, went viral all over the world. Millions of viewers changed their minds about sharks.

Despite this, most humans, especially those who will never get close to sharks, remain convinced they are man-eaters and that it is a good thing to exterminate them.

However, for those like me who have sojourned with them for a while in their territory, this fear, based as it is in fantasy, is incomprehensible.

More seriously, it is unacceptable to see, in the midst of a general indifference, the dramatic collapse in numbers of all shark species and the virtual disappearance of some of them.

So, this book pleads the case for Lady Mystery, and for all sharks, for all the wild things who will never be able to speak up, for all those who are different and whom we fear out of ignorance. But it also pleads our own case, because we feel, as Romain Gary put it so eloquently in his 'Letter to an Elephant',[3] that it is our humanity that we are massacring when we wipe out Lady Mystery's freedom in the wild.

To be a good advocate, you need to be well informed. You have to get inside the heads of those you are defending. You have to be willing to lose yourself in the world of the other through their senses, you have to try to approach their *Umwelt*.[4]

In addition to the hundreds of dives I have made with all kinds of sharks, in all the seas of the world, this book takes note of the most recent scientific research on their biology and their exceptional sensory system. It focuses on the discoveries in ethology and neurobiology which, incredible as it may seem, are able to determine each shark's individual personality.

This book also focuses on the men and women who dive with sharks and are committed to defending them. It shows how they have broken through the barrier of prejudice to discover the true nature of sharks. The hope is that the reader will be encouraged to take the same approach so that everyone can form their own opinion and no longer submit to the preconceived ideas that, tirelessly repeated by people who have never seen a shark, have become unquestioned commonplaces.

For in the end, while we humans progressively colonize a little more of the territory of others, without care for the rules of their ecosystems, contemptuous of their existence and their signals, thus multiplying the risk of dramatic confrontations, it is indeed a question of finding a diplomatic way, as Baptiste Morizot calls it, that will allow us to live in peace.[5]

This book is dedicated to our oceanic cousins, and it is also a reflection on our relationship to the world and to otherness in general. The shark, as a symbol of the 'wildness' outside of our rules, that frightens us because of our ignorance, a symbol of all that is useless and getting in our way, extending to all those who are different in their ways of life, their traditions, religions and cultures. In this sense, getting to know wild animals in order to try to reach a diplomatic consensus, could offer good lessons for living in society.

Finding out what each shark's singularity is forces us to ask questions about its status as a non-human person, and consequently its right to exist. From there, more globally, we can rethink our place at the heart of the earthly ecosystem, alongside our wild companions. All living beings share the planet as our common good.

1

A Matter of Misunderstanding: From Pliny to Disney

The axe comes down on the body, which twists and turns. It strikes again and again. An incredible violence that seems to express all the repugnance humanity has towards such a vile beast: the shark! It is 1954, in the middle of the Indian Ocean, on the deck of the *Calypso*, Captain Cousteau's ship; he concludes with a terse: 'Sailors the world over detest sharks . . . But for us divers, the shark is our mortal enemy.'[1]

Thirty-five years later, aboard the same *Calypso*, in the same Indian Ocean, we set sail, off the shipping lanes, towards the mysterious Andaman archipelago. We aim to make an inventory of the marine fauna of this region where no one has ever dived and which, in 1989, was still free of industrial fishing. But our secret hope, after having scoured the Caribbean Sea, after having searched the Pacific Ocean, from the Marquesas Islands to the Great Barrier Reef, is to finally find a virgin ecosystem, full of shark communities. A place where sharks abound, a place so famous that Jules Verne himself refers to it in *Twenty Thousand Leagues under the Sea*: 'I know well that in certain countries, particularly in the Andaman Islands, the negroes never hesitate to attack them with a dagger in one hand and a running noose in the other; but I also know that few who affront those creatures ever return alive.'[2]

By 1989, the world had definitely changed.[3] The exploitation of 'marine resources' was at its peak. More than ninety million tonnes of fish were shipped worldwide, a critical threshold that would never be reached again.[4] The shark is no longer the enemy; it is in danger. And Cousteau, whose exploration of the marine world had turned him into an environmental protector, wanted to sound a warning about the massive and rapid disappearance of sharks. Perhaps, inwardly, he wanted to repair the damage his film *The Silent World* had done to them.

The Andaman Islands sharks

On 4 April 1989, *Calypso* dropped anchor at 11° 8′ north latitude, 93° 31′ east longitude, on Flat Rock, Invisible Bank. Five o'clock in the morning. Camera, writing slate, sample tubes, everything is ready for the first reconnaissance dive. Dawn is just breaking. The sky and the sea merge into a greyish uniformity. The 'Invisible Bank' is formed out of a skull-like volcanic rock crowned with foam at water level. It is low tide. We tumble down under the surface. With no time to turn around, to take our first breath, there are three silvertip sharks (*Carcharhinus albimarginatus*) and two large grey reef sharks (*Carcharhinus amblyrhynchos*) coming straight towards us. How did I identify them without seeing them properly? Their bodies are barely visible. Only the bright fin edges are dancing in the obscure depths. Opalescent flames circling quietly in the distance, then they wheel and suddenly shoot off.

The sea seems to be giving birth. Sharks coming into the world: first the black crescent of the mouth which contrasts with the paleness of the snout, then the powerful roundness of the body, stabilized in the dense water by the pectoral and dorsal fins. The eye is perfectly round. A golden iris with a vertical anthracite pupil. It stares without letting go. The five gill slits, like exclamation marks. Then the muscles playing under the skin that is both granular and silky. A mass of energy, contained, ready to explode. Supreme power. Finally, the whip of the tail, like a white flag that leaves you spinning.

We reach the bottom. We wedge ourselves between a coral mass and a red sea fan. The biggest shark comes up to us, a large female grey reef shark. She measures at least 2.8 metres, maybe three metres, a giant for a species whose biggest individuals rarely exceed two and a half metres. This is a sure sign, no fishing in these parts. It is a pristine ecosystem. Indeed, these rare mature giants are the first to disappear as soon as fishing begins. And they are never replaced, because the harvesting rate is such that the young ones no longer have time to grow older and larger. This matriarch shows deep bites on her flanks and a large tear on her left fin, traces of her numerous couplings. This one scarred female tells the story of a world, of a former Earth, populated by such noble savages, prodigiously enormous beings who had grown old unmolested. This makes me forget the notes scribbled on my slate and my samples. Dull figures and words cannot convey such an impact.

Cousteau and Lord *Longimanus*

Returning from this survey, the decision was made to use the protective cage, given the abundance of large sharks. In these unexplored waters, Cousteau did not want to take any risks. He had been cautious since his misadventure with a big oceanic whitetip shark (*Carcharhinus longimanus*)[5] in 1948 off the Cape Verde Islands. Memory does not fade away so easily. It left a deep impression on him, so much so that he, often discrete about his past adventures, told us the story many times around the *Calypso*'s dining table, and described it in several books, with various embellishments over time:

> We had scarcely entered the water and were only fifteen or twenty feet below the surface when we saw Lord Longimanus ... He resembled none of the sharks we had met before. ... Confident – too confident – in ourselves, we dropped the line that still linked us with the ship [the *Elie Monnier*] and swam straight toward him. His squat, gray-brown silhouette was sharply etched against the clear blue of the water. His head was very round and very large, his pectoral fins enormous and his dorsal fin rounded at its extremities. ... It was time – much too long a time – before we realized that the Lord of the Long Arms was drawing us with him into the distance, but was not in the least afraid of our approach. As soon as we realized this, we were seized with an almost paralytic fear and wanted nothing more than to return to our ship. But it was too late ... Two blue sharks, very large but classic in form, came to join our *longimanus* and then the three squali began to dance around us, in a gradually narrowing circle. For twenty seemingly interminable minutes, the three sharks, prudently but resolutely, tempted a bite at us each time we turned our back on them or each time one of us went up to the surface to signal – in vain – to our far-off ship. Miraculously, the gig which the captain of the *Elie Monnier* had put overboard to look for us found us and saved us from imminent death. Shortly before we were hauled from the water I had arrived at the point of smashing my camera against the head of the *longimanus*, in the forlorn hope of warding off his attack and gaining a little time.[6]

Final exit from the shark cage

Even though we were not in the same situation – we didn't have three kilometres of water under our flippers, only fifteen metres – we did not want to argue about J.-Y. C.'s (Jacques-Yves Cousteau's) instructions or put ourselves in a difficult situation simply out of bravado. So, at 9.30 a.m. the shark cage was hanging from the end of the crane behind *Calypso*. It looked a bit like one of those papier mâché Mardi Gras decorations that are burned in public to mark the end of winter. It was a disturbing impression. Those steel bars, carefully painted yellow, behind which the pioneers of underwater exploration took refuge in 'shark-infested' waters, seemed quite useless at that moment. We had already had so many encounters with sharks that we had changed our minds about these unloved creatures. At the same time, this cage, the stuff of dreams of generations of divers and millions of television viewers, which had given me comfort in my childhood, linked me fraternally to my predecessors, those marvellous heroes of *Calypso*'s earlier odyssey.

Solemnly lowered in the Zodiac, the cage was tipped into the sea at the dive site, a few hundred metres further on. And what everyone predicted happened: the great whitetip sharks kept their distance, ceding the place to the many colourful small fry that are the soul of the reef. Only a few whitetip reef sharks (*Triaenodon obesus*) and a tawny nurse (*Nebrius ferrugineus*) came close enough to bump into Didier Noirot's camera. No need to take refuge behind bars!

That was the last time the shark cage was used. Before day's end it was dismantled and stowed, never again to bring a yellow glow to the Cousteau films; that special hue, that little something that made us *Calypso*'s intrepid frogmen. A page in the history of our relationship with sharks had been turned.

A peaceful dusk among the sharks

It was a frustrating day, so I dived again at dusk, this time alone. The sleek unicorn fish (*Naso hexacanthus*) and banana fusiliers (*Pterocaesio pisang*) that enliven the reef were already lying in crevices. I found the big whitetip sharks again. A dozen or so on the hunt. As in the morning, I could hardly make out the grey bodies that were blending in with the masses of coral reefs. But, despite the darkness – or perhaps because of it – the white tips edging their fins seemed luminescent. I found this ballet of white flashes hypnotizing and soothing, transporting me to a powerful, wild, former world, like a time traveller in communion with all the other living creatures around me, not only the sharks, but the fish and the coral. I felt *alive*, a part of that great incomprehensible *wholeness* – in the sense that it is totally beyond us – which is the mystery of life. Being part of the world of the animals, a world of raw meaning, without complex reasoning, intuitive, primordial. I imagined myself as a shark, or perhaps like those ambivalent shamanic creatures who bridge the gap between humans and Others, as a lycanthrope, a cynocephalus, half-man half-beast. I was overwhelmed by a deep desire to speak for the sharks, to be their conduit, their ambassador to my fellow man. I wanted to shout out loud about the peace they can bring.

'Peace', 'calm', 'serenity', these are the words that best describe these dives among our shark cousins. Writing now in 2021, thirty years later, I feel even more how privileged I was to have these precious encounters, the happiness of these rare moments.

Fig. 1 The final shark cage dive in the Andaman Islands as we dreamed it would be, surrounded by whitetip sharks.

An animal familiar to all, but which very few have seen

What a contrast with the fear that the mere mention of sharks arouses. Why is it so deeply rooted in our collective imagination, even among people who have never seen the sea, let alone sharks? Fear and rejection seem to be universal, whether among city dwellers in overpopulated megacities or cattle farmers in the American heartland.

In 1970, when he did a rerun of a television series, Cousteau did not hesitate; he chose to make a documentary devoted solely to sharks. He understood that, more than any other subject – sunken treasures and cities, dolphins and whales – sharks would fire up the viewers of his *Odyssey*:[7]

> The first film in the series was scheduled to be the one most likely to intrigue and attract the attention of the viewers', and what maritime subject is more fascinating to everyone than the shark? It is a legendary animal, known to all, even to those who live far from the sea.[8]

And that is the heart of the matter: an *animal-familiar-to-us-all*, but which we know no better than the dinosaurs that disappeared sixty-five million years ago.

Where does this familiarity with an unfamiliar animal come from? In the West, there is no trace of sharks in our legends or in our myths, which are, on the other hand, populated by wolves, bears, witches and dragons. Except perhaps in one of the many versions of the Greek legend of Lamia, daughter of Poseidon, who was turned into a shark by Zeus so that she could avenge the massacre of her children by devouring those of other peoples.[9]

The realism of the early naturalists

Unlike the fantastic accounts usually expected of ancient writers, the ancients' description of sharks is more realistic than poetic. In his *History of Animals*, written in 343 BCE, Aristotle already recognized several species: the great white, the fox, the smooth hammerhead (*Sphyrna zygaena*), the blue shark, the spiny dogfish (*Squalus acanthias*), the dogfish and the tope (*Galeorhinus galeus*).[10]

Three hundred years later, Pliny the Elder took up Aristotle's writings and provided very relevant information on the reproduction of 'cartilaginous fish . . . While the (bony) fishes are oviparous, these (the selachians) are viviparous like the cetaceans'.[11] Pliny focuses on the threat that sharks pose to sponge fishermen:

A multitude of sharks infest the seas[12] where the sponges are, to the great danger of divers . . . [. . .] They attack the groins, the heels and all the white parts of the body. The only resource is to go in front of them and take the offensive. Indeed, they are as much afraid of man as they are of him. Underwater the game is equal, but on the surface of the water the danger is imminent, the diver loses the advantage of facing the shark as soon as he tries to surface. His only hope is in his companions.[13]

Centuries go by without a word on sharks

Little was added to our knowledge for centuries. Descriptions of marine animals, and especially sharks, were rare, particularly since scientists, philosophers and scholars were not sailors. They did not venture out onto the ocean where the underworld lies. They leave it to a few intrepid men, not quite human, not quite alive, nor even dead, to be bold enough to venture out onto the waves. Anacharsis, the philosopher, was astonished by the thinness of the hull of ships separating 'the world of the living from that of the dead'.[14,15]

Worse still, knowledge of the marine world regressed dramatically in the Middle Ages. The Earth, which Anaxi Mandra of Miletus[16] already knew to be spherical in the fifth century BCE and whose circumference had been calculated by Eratosthenes[17] at the beginning of the second century BCE, became flat once again!

It was just a disc at that time, a central land area surrounded by the ocean, and at the border people fall into the abyss.[18] Needless to say, in this context, history with a 'human-terrestrials' bias is written without sharks being properly described or even mentioned in the inventory of monsters.

By 1539 nothing had changed, but after twelve years of hard work, the Swedish archbishop Olaus Magnus offered Renaissance scientists his 'Carta Marina', a representation that brought together the knowledge that Westerners had of the marine world. Sharks did not even figure in his incredible bestiary, a dreadful world where mermaids, sea pigs, whales, unicorns, sea snakes, sea bishops, krakens and other dragons are milling about.

Even the famous naturalist Pierre Belon maintained the mixture of real and legendary in his *Histoire naturelle des estranges poissons marins*,[19] written in 1551. There is a detailed description of the 'monk-like sea monster' as well as a fairly precise description of the sharks brought back by fishermen. In his magnificent 1577 *Whale Book*, Adriaen Coenen presented realistic paintings of sharks brought in to Dutch ports: the spurdog, the school shark, the hammerhead shark and the angular roughshark (*Oxynotus centrina*). But he too continued to pay a good deal of attention to sea monks and mermaids, still the most dangerous of monsters, along with the sperm whale and the kraken.[20]

Only Guillaume Rondelet, in his book *De Piscibus Marinis*, argued forcefully in 1554 that the Leviathan that swallowed Jonah was not a whale but a great shark, the lamia. 'This fish is very greedy. It devours men whole, as we know from experience. In Nice and Marseilles, lamia were once caught, and in their stomachs they found a whole man with his armour on ...'[21] Is Rondelet referring to the legend of Lamia? In any case, this name was used by Mediterranean fishermen to designate the great white shark until the twentieth century.

Shark and wolf

At the same time, on the other side of the Earth, in the heart of the immense Pacific Ocean, the Polynesians worshipped sharks. Setting out from Melanesia in sailing canoes that carried entire clans, these exceptional sailors, experts in the art of orienting themselves by the stars, sailed for weeks on end without any land in sight. 'Mother Earth' for them is the ocean. Their mythology and stories are filled with sea creatures. Their gods are sperm whales, whales, turtles and, of course, sharks. These protective deities guide the sailors on their perilous journeys. Kamohoalii reigns over the pantheon of shark gods. He can take human form. Assisted by Kane-i-kokala and Ka'ahupahau, a human-born shark goddess, he protects sailors and rescues those shipwrecked.[22]

The Maori legend of the impossible love between Kawariki, the daughter of the Matakite sorcerer, and Tutira, the low-born slave transformed into a shark, is reminiscent of the story of Romeo and Juliet. But, as a seafaring people cannot get angry with the sea, the story has a happy ending thanks to Hinemoana, the goddess of the Ocean.[23]

As one might expect, in Europe, Asia, Africa, among all the 'landlocked' peoples, the shark, just like other sea monsters, does not feature in popular stories. Tigers, lions, turtles and wolves reign supreme. In Scandinavian mythology, the wolf, *Fenrir*, is an evil one, terrorizing the people, the knights and even the gods. For the Iroquois and Sioux, the wolf is a benefactor, guiding the souls of warriors across the plains of the Great Spirit. He is the father of the Mongolian people whose kings, led by Genghis Khan, are descended from the Blue Wolf, *Börtea-Chino*, symbol of Heaven. A she-wolf even presides over the foundation of Rome by suckling Romulus and Remus. The wolf is omnipresent, and a great master in extraordinary legends.

The beast of Gévaudan

France had nearly 20,000 wolves before the nineteenth century, so the story of the Beast of Gévaudan spread easily, terrorizing the whole country. The national and even the international press reported each attack extensively. For the first time, the media played a role in the creation of a supernatural being, installing the beast in everyone's mind. The wolf emerged from folklore and entered the pantheon of diabolical creatures.

Marine creatures did not benefit from such public exposure. Notwithstanding a few frightening or heroic tales told by sailors, sharks did not interest a France of city dwellers and peasants. And it wasn't John Singleton Copley's 1778 painting, *Watson and the Shark*, that changed this, especially as this work – in which a shark attempts to devour Brook Watson (who was actually bitten and later became mayor of London) – remains exceptional among marine paintings.[24]

Paradoxically, in the nineteenth century, Mediterranean tuna fishermen appreciated the help of sharks. The naturalist Marcel de Scrres has the following description:

We also see mackerel eating sardines just like tuna eat mackerel. The tuna themselves are in turn devoured by the sharks, which pursue them with such relentlessness and a kind of fury that they prefer to beach themselves on the coast rather than suffer the cruel death that awaits them under the sharp teeth of these tigers of the seas, with their insatiable appetites. The fishermen take advantage of the terror that the sharks inspire in the tuna to pick them up during the day. These facts are so well known to the fishermen of the Mediterranean coasts that the appearance of sharks is, in their eyes, a good sign that tuna are running.[25]

The literary debuts of the sea monsters

Thanks to the talent of some of the major writers of the nineteenth century, Westerners discovered the amazing bestiary, the real one, populating the ocean. The sperm whale, Moby Dick, was brought to the fore by Herman Melville, and the giant octopus was celebrated by Victor Hugo: 'The devil-fish has no muscular organization, no menacing cry, no breastplate, no horn, no dart, no claw, no tail with which to hold or bruise; no cutting fins, or wings with nails, no prickles, no sword, no electric discharge, no poison, no talons, no beak, no teeth. Yet he is of all creatures the most formidably armed. What, then, is the devil-fish? It is the sea vampire.'[26]

Barely three years later, in 1869, Jules Verne, who, with his monumental *Twenty Thousand Leagues under the Sea*, wanted to take readers on an adventure and offer them up-to-date knowledge as well, was to further reinforce the public's aversion to the octopus: 'a horrible monster worthy to figure in the legends of the marvellous. It was an immense cuttlefish, being eight yards long ... watching us with its enormous staring green eyes. Its eight arms, or rather feet, fixed to its head, that have given the name of cephalopod to these animals . . . were twisted like the furies' hair.'[27]

Appearing in the nineteenth-century bestiary . . .

The cover of the original book, published by Hetzel, shows two giant octopuses, a whale, a narwhal, an eel and two divers, but no sharks. Would Jules Verne have left them out of his marine bestiary too? No, on the contrary. He takes advantage of the presence on board the *Nautilus* of the learned naturalist, Conseil, to instruct the reader: '. . . the selachians, with gills similar to those of the cyclostomes, but whose lower jaw is mobile. This order, which is the most important of the class, includes two typical families: the rays and the sharks.' Nemo, mocking, does not let the naturalist glory in his bookish knowledge. He suggests that he go hunting for sharks in their natural element, underwater:

> Now, if you were invited to hunt the bear in the mountains of Switzerland, what would you say? 'Very well! to-morrow we will go and hunt the bear.' If you were asked to hunt the lion in the plains of Atlas, or the tiger in the Indian jungles, what would you say? 'Ha! ha! it seems we are going to hunt the tiger or the lion!' But when you are invited to hunt the shark in its natural element, you would perhaps reflect before accepting the invitation.

Shortly afterwards, wearing a diving suit looking like a suit of armour, Conseil left the *Nautilus* alongside Nemo: 'My blood froze in my veins as I recognized two formidable sharks which threatened us. It was a couple of tintoreas, terrible creatures, with enormous tails and a dull glassy stare, the phosphorescent matter ejected from holes pierced around the muzzle. Monstrous brutes! which would crush a whole man in their iron jaws.' A few pages later, they witness an attack on a pearl fisherman: 'The voracious creature shot towards the Indian, who threw himself on one side to avoid the shark's fins; but not its tail, for it struck his chest and stretched him on the ground.'

These popular epics excite readers but do not terrify them. The horrible descriptions and Homeric battles anoint heroes, Nemo and his *Nautilus*, or embattled pairs, Ahab and Moby Dick, Gilliatt and the octopus, as in the past Hercules was glorified by the Hydra of Lerna, Theseus by the Minotaur and Saint George by the Dragon. These demigods are not close enough to people to worry them too much. Sea monsters are still too superior to take the place of the wolf. On the other hand, the Beast of Gévaudan left a deep and lasting impression on people's minds because it attacked anonymous people; anyone could be gobbled up.

The wolf's successor

At the dawn of the twenty-first century, untouched forests and swamps were being redeveloped. Wild animals were banished from domesticated areas and no longer threatened humans. Wolves, bears and tigers were hunted down and obliterated. Land-based monsters no longer frightened people, not even children believed in them any more. Substitute monsters needed to be found, real ones, ones that really threaten people.

Sensationalist newspapers[28] were regularly publishing front-page pictures of adventurers or convicts crammed onto makeshift rafts at the mercy of threatening sharks. But the fate of these unfortunates did not really touch the average person. Country folk and city dwellers alike were still indifferent. In order to create more of a psychosis, such creatures had to

SUR LES CÔTES DE LA GUYANE Forçats évadés aux prises avec une bande de requins

Fig. 2 Escaped convicts on a raft surrounded by sharks. *Le Petit Journal Illustré*, 20 May 1906.

lunge out of nowhere and interrupt the everyday lives of ordinary people.

In the world of ordinary people

And that's exactly what happened in July 1916, on the New Jersey coast, south of New York. The bathing season was in full swing, in the full knowledge that sharks were present along the coast. Thousands of carefree bathers were enjoying a refreshing swim with their families. Never before had there been any remarkable incidents. Suddenly, in less than a fortnight, sharks killed four people and savaged another.

On Saturday 1 July, 23-year-old Charles Epting Vansant, who had come with his family from Philadelphia, was playing with his dog a few metres from the shore. He was bitten, but the rescuers arrived too late to stop the bleeding. Five days later, Charles Bruder, 27, an employee of a luxury hotel, was bitten a hundred metres from the beach. He also died of his haemorrhages. Then it was the turn of 11-year-old Lester Stillwell, 24-year-old Watson Stanley Fisher and 14-year-old Joseph Dunn. Suddenly, it's a case of you, me, anyone; Mr Everyman who is threatened by the death lurking beneath the ocean's surface. The accidents triggered a wave of panic the likes of which the United States had never seen. The authorities organized fishing operations to eradicate the man-eating sharks.

Extremist media coverage thus overturned public opinion, which had previously thought sharks to be harmless. They were now seen as killing machines. From now on political cartoonists in the press use the image of the shark to symbolize horror and perversion.[29]

Thirty years later, the sinking of the cruiser USS *Indianapolis* would finally seal the 'shark–evil' association. On 30 July 1945, the US Navy vessel was torpedoed by a Japanese submarine. The ship sank in twelve minutes, drowning 300 of the 1,197 crew members. The survivors drifted in the open sea for four days and nights. Only 300 survived. Although their testimonies exonerated the sharks, which only devoured about fifty sailors who had already died of hypothermia and dehydration, the navy's accounts and the media denounced the culprits: the sharks! Yes, the sharks, they were the cause of the horror, almost of the whole shipwreck. The Japanese torpedoes and the carnage of the war in the Pacific faded into the background.

A cultural phenomenon

From then on, the rise of sharks to the pinnacle of monstrosity only accelerated. Any marine odyssey or adventure would have to have the heroes dealing with sharks. Even Tintin, on his hunt for Red Rackham's treasure, is shown on the cover in a shark-shaped submarine.[30]

In 1952, *The Old Man and the Sea*, by Nobel Prize winner Ernest Hemingway, elevated sharks to the literary pantheon. The old man's epic battle with his giant marlin, and then his desperate battle with the sharks, hit the nail on the head: 'The shark closed fast astern and when he hit the fish the old man saw his mouth open and his strange eyes and the clicking chop of the teeth as he drove forward in the meat just above the tail. The shark's head was out of water and his back was coming out ... There was only the heavy sharp blue head and the big eyes and the clicking, thrusting all-swallowing jaws.'[31]

A shark in every home

The fear of sharks emerged from the convergence of three factors: the discovery of the ocean floor, the disappearance of other wild animals and media globalization.

In 1956, the phenomenal success of the film *Silent World*, which won the Palme d'Or at Cannes and an Oscar in Hollywood, drew the whole world into the depths of the unknown ocean. The scene of the shark massacre, which is so shocking today, filled millions of spectators with delight and fear as they discovered the man-eating beast for the first time. Sharks arrived in the nick of time to replace wolves and other wild animals that had been wiped out by man.

The public was fascinated by the conquest of this new territory: 'The Americans are going to conquer the Moon, I am going to conquer the sea', said Cousteau. *Calypso* travelled the oceans. The dives multiplied and so did encounters with sharks. But above all, for the first time in history, they were popularized on a global scale. Cousteau, *Calypso* and the sharks benefited from the huge expansion of television, which entered every home and, without competition, invaded the world. Each episode of the Cousteau odyssey was seen by several hundred million viewers. Admired, feared and disliked, sharks become stars anyway. They even invaded shop displays with other beach toys.

I remember the magnificent inflatable shark that my mother gave me in the summer of 1958 for a holiday by the sea. It was bigger than I was and it took the fancy of André Deval, *Dauphiné libéré* photographer, who published the photo with this headline: 'With a blue shark in my arms . . . François winked at his mother, as if to say: "This is the playmate I really want to go to sea with."'

The media overkill increased the pressure. And when *Calypso* left in 1967 to film the sharks, even the most seasoned divers feared meeting them: 'On the night of our departure . . . my head was filled with thoughts of that fabulous animal, the redoubtable man-eater, and of the metallic beauty and invincible strength of that incomprehensible monster – the shark. No creatures of the sea . . . have inspired irrational fear in me: none, that is, except the shark', Philippe Cousteau wrote in his diary.[32]

And the shark became *Jaws*

Finally, three bestsellers definitively drove the point home. In 1972, the documentary film *Blue Water, White Death* by James

Fig. 3 *Le Dauphiné Libéré*, 18 July 1958. The author 'with a blue shark in his arms'.

Lipscomb and Peter Gimbel received very favourable reviews: 'Two times fifteen minutes of fascinating images: the evolution of scuba divers in the midst of hungry sharks, the attack of a *Carcharodon carcharias* or white shark (length: 10 metres; weight: 3 tonnes)[33] against an aluminium cage in which a man is locked up. It certainly took a lot of courage and nerve to make this documentary film about the "killers of the seas"'.[34] Then, in 1974, Peter Benchley's novel *Jaws*, which sold over nine million copies. And the following year, Steven Spielberg's film of the same name stayed at the top of the US box office for forty-four weeks, one of the biggest financial successes of all time.

An untruth repeated a thousand times becomes a truth; repeated ten thousand times, it becomes so paradigmatic that many scientists no longer question it.

When, in 2006, for the film *Oceans*, we wanted to shoot a scene showing that a diver could swim in harmony with a great white shark, it was the American scientists who most vigorously opposed a cage-free approach. However, they had no experience to back up their fears. They were only arguing a prioris against the very real experience of our free dives with South African white sharks, guided by Andre Hartman, in July 2000.[35] Rumour was stronger than fact, and habits were resistant to new hypotheses: the great white shark remained a killing machine. At best one could observe it from behind the bars of a cage. Only a few shark researchers, such as Eric Clua, maintained that it was necessary to dive with sharks to study them.

Even today, many behaviourist[36] scientists consider animals, especially cold-blooded ones, as automatons that respond in a standardized way to environmental stimuli in order to satisfy their needs: to feed, to avoid predators and to reproduce.[37]

Nemo, Shark Tale: The 'Disneyfication' of the world

In contrast to this animal-as-automaton theory, the 'human sharks' in animated films such as *Finding Nemo* (2003)[38] and *Shark Tale* (2004) are just as caricatural. The sharks are kind, sometimes even vegetarian. They are compassionate and know all about the problems of small-town Americans! As ridiculous as the concept is of the Machiavellian shark as proposed by *Jaws* or *Deep Blue Sea* (1999), this anthropomorphic representation is more disturbing. It even appeals to some shark 'defenders' because of the work it does to rehabilitate sharks by eliminating their cold-blooded-killer image. In *Shark Tale*, there is always its counterpart, *Jaws*, with numerous references. Unfortunately, this representation reinforces misunderstandings and further increases the gap between reality – life in the wild – and 'out-of-touch' city dwellers, who don't even know about cows, chickens or nature, and whose new monsters are Transformers robots. More seriously, to the extent that some of the biological observations are truthfully filling in character and setting, confusion is created in the minds of the public, who see the films as educational. The most perverse argument is that these films allow children who do not have access to the sea to find out about the creatures that inhabit it. This argument is also used for Disneyland theme parks, which invite people to become immersed in the heart of a 'more real than reality' nature, and in Marineland, which shows orcas and dolphins playing with balloons. This mystification reinforces the idea that a species is defined by its morphology and can exist outside its ecosystem. It thus legitimizes *ex situ* conservation and the 'Noah's Ark' concept that forgets that it is the relationship with other living beings and the physical environment that defines a living being.[39] The ambiguity arises from the fact that most people think that the 'appearance' of the species is sufficient to know the species.

Worse, this representation of a sanitized nature contributes to the idea that wildlife can be advantageously replaced by a safe and accessible representation for all. It establishes the archetype of a 'nature backdrop' populated by animals intended to satisfy our whims. In this context, Lenny, the vegetarian great white shark in the film *Shark Tale*, fits in perfectly, just as teddy bears found their niche making children forget that *Ursus arctos* once shared the same territory as us humans.

And man created the virtual shark in his image

Acting both Dr Jekyll and Mr Hyde parts, the great white (vegan in animated films) retains his role as a Machiavellian killer in over twenty horror films, laughable remakes of *Jaws*. He is the embodiment of all 536 current species and brings all our fantasies to life. So why bother with the other species, those also-rans which, in our 'utilitarian' management of the world, are much less effective than the great white shark of the cinema? This symbolic shark fits perfectly with the emerging concept of the 'functionalization of living things', which aims to preserve living things because of the 'services they render', and not the living beings themselves, even if it means sacrificing the diversity of species because some might be 'redundant'. This forgets that the strength of nature and its resilience lie in this redundancy and extreme diversity, which are the result of the natural diversification (mistakenly called 'natural selection')[40] that has been at work since the dawn of the world.

Finally, this iconic simplification satisfies our condescending desire to be 'nature's managers', symbolized by Noah – champion of the culture/nature dualism – in that it definitively puts the Human Being in the position of the world's chief manager.

In an increasingly virtual society, where the *image* is a tsunami that overwhelms all reality, man becomes the great creator. The eleven billion views (yes, you read that right: eleven billion) of the children's video *Baby Shark* on YouTube are a testament to the ability to erase reality on a scale never before achieved.[41]

So, let's delve into the roots of the diversity of the 536 species of sharks to get a better idea of the extent of their history and their co-evolution with other living beings.

2

Shark? What Shark?

'If you want to dive with sharks, you have to go to Australia, that's where they filmed *Jaws*!' I was told this any number of times when I was preparing expeditions on the *Calypso*. Well, we finally got there on 4 November 1987, and the *Calypso* dropped anchor at 11° 35' 50" south latitude and 144° 2' 00" east longitude, south of Raine Island, the world's largest sanctuary for green turtles (*Chelonia mydas*). The moon is full. Thousands of these large sea turtles are emerging from the ocean onto a beach barely two kilometres long to lay their eggs in the warm sand. It's a gruelling process for these marine creatures, but a bonanza for the big sharks, who wait at the edge of the reef for the females to return to the sea, exhausted by their laborious night. It is a great opportunity for our team, who have high hopes for these dives in the 'shark-infested' Coral Sea.

As soon as we put our heads under, we can make out the silhouette of a tiger shark (*Galeocerdo cuvier*) in the distance, which has come for the feast. But it disappears as we approach. As we dive, we are in contact with turtles swimming in all directions, jostling each other, mating and biting. But no big sharks. We were about to return empty-handed when I saw a strange creature on the bottom, between two enormous porite coral colonies (*Porites lutea*). It is snake-like, brown with black spots, and barely a metre long, but it is walking on its fins! Yes, on four fins, moving one after the other, a bit like a lizard. It uses its pectoral and pelvic fins as legs. I exhale slowly so as to sink to the level of this strange creature. It continues its reptilian movement, without a worry in the world until it touches my mask! It is a small shark, a crab and worm eater, called the epaulette (*Hemiscyllium ocellatum*), whose strange 'gait' no one had ever observed. Thirty years later, a genetic study showed that this shark and one of its cousins, the hedgehog ray, or little skate (*Leucoraja erinacea*), had inherited the gene that enables them to walk, a gene that was already present in a common ancestor 420 million years ago.[1]

Thrilled by this encounter, we continue exploring the reef, discretely observing the turtles and coral sharks that loom up ahead. I adopt the exploratory tactic known as the 'helicopter survey', three metres above the reef. At this height I can catch any action taking place in the open water, but I am close enough to the bottom to see exactly what is hidden there. I skim over a field of coral debris, a vast heap of broken acroporid coral branches, gastropod shells, *Halimeda* calcareous algae, where all sorts of invertebrates, sea urchins and sea cucumbers swarm. Nothing that really catches the eye. And yet, I am sure I saw something. I don't know what. Something bilaterally symmetrical, totally ectopic[2] in this chaos. I call my team members; I turn back, looking around, not knowing for what exactly. And suddenly I see him, as if it were obvious. First his eyes and his small dorsal fins that betray a silhouette with indefinable contours. The flat body, variegated with light and dark marbling, seems to be part of the background. Finally, the head, adorned with skin expansions so tortured that they look like algae. Perfect camouflage. It is an ornate wobbegong (*Orectolobus ornatus*), a great hunter on the prowl.

Sharks: a shape to suit every taste!

Not all sharks have the size and appearance of the shark of our nightmares. Some, such as the carpet shark or the angelshark (*Squatina squatina*), live on the bottom and are so flat that we would be happy to classify them as rays.

On the other hand, the bowmouth guitarfish (*Rhina ancylostoma*), also known as the shark ray or mud skate, has such a rounded back and such high dorsal and caudal fins that it would be better classified as a shark! And what about the smalltooth sawfish (*Pristis pectinata*), which is classified as a ray, while the common sawshark (*Pristiophorus cirratus*), which could be its twin, is a shark! Additionally, in Mexico, researchers have just discovered, in Upper Cretaceous deposits, a planktivorous shark related to the white shark (*Aquilolamna milarcae*),[3] whose pectoral fins are reminiscent of the 'wings' of the oceanic manta ray (*Mobula birostris*)! Skates and sharks are such close cousins that only the position of the gill slits can tell them apart. Gill slits that open on the ventral side: ray. Gill slits that open on the sides, at the back of the head: shark.

The bull shark, with lateral gill slits, and the eagle ray, with ventral gill slits, are close cousins, although their morphology and swimming patterns are quite different.

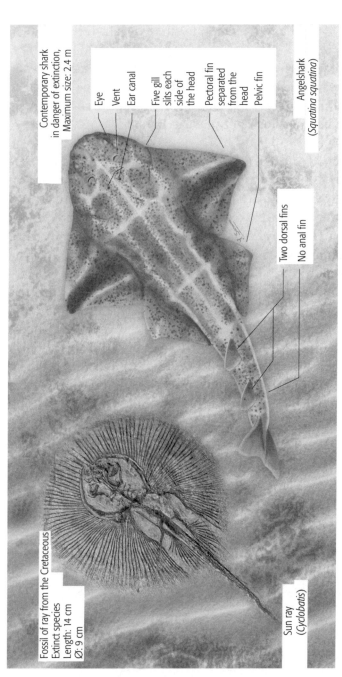

Contemporary shark in danger of extinction, Maximum size: 2.4 m

Eye
Vent
Ear canal

Five gill slits each side of the head

Pectoral fin separated from the head

Pelvic fin

Angelshark (*Squatina squatina*)

Two dorsal fins
No anal fin

Fossil of ray from the Cretaceous
Extinct species
Length: 14 cm
Ø: 9 cm

Sun ray (*Cyclobatis*)

Fig. 4 Fossil of the sun ray (*Cyclobatis* sp.), extinct since the end of the Cretaceous period, and the angelshark, whose flattened shape is reminiscent of a ray. The species is critically endangered.

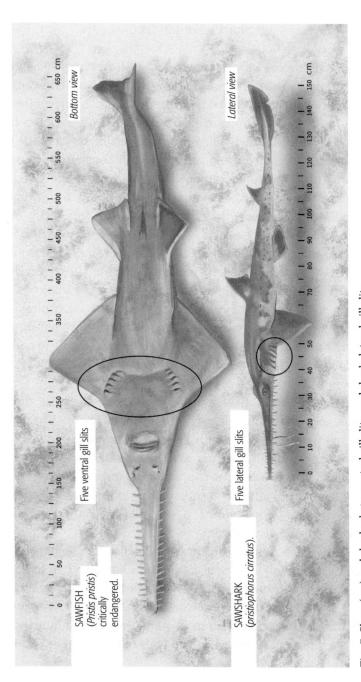

Fig. 5 Skate (ray) and shark: skate, ventral gill slits, and shark, lateral gill slits.

Bottom view

650 cm

Lateral view

150 cm

Five ventral gill slits

SAWFISH
(*Pristis pristis*)
critically
endangered.

Five lateral gill slits

SAWSHARK
(*pristiophorus cirratus*).

Sharks are not fish!

Rays are therefore only flattened sharks. In fact, their flexible cartilaginous skeleton classes them with the chondrichthyans. And it definitely distances them from all other vertebrates, even ordinary fish (grouper, carp, salmon) whose skeleton is ossified (osteichthyans).[4] Pliny the Elder, relying on Aristotle's discoveries, already distinguished fish with bones from those with cartilage, which he grouped together under the Greek name *selaxe*: selacians, or cartilaginous.[5] (See Fig. 6 showing vertebrate phylogeny.)

Another morphological detail distinguishes ordinary fish from sharks: the cavity in which their gills are housed. The gill cavity of fish is hidden by a lid, whereas the gill cavity of sharks has five open gill slits, or almost, because sharks do as they please, or rather as their gills please; most of them have five gill slits, others have six or even seven. These differences make it possible to classify the large families that exist today (see Fig. 14 on page 56 with the orders of sharks).

Rays and sharks have another feature in common that makes them even more different from fish: their skin. It is not covered with scales, but with millions of microscopic teeth, consisting of pulp and dentine protected by an enamel coating. This skin, armed with denticles that are regularly renewed, is so strong and rough that it was once used as an abrasive by cabinetmakers. Once tanned, it was sold by leatherworkers under the name of 'shagreen'.

Sharks are definitely not fish. And if we risk a comparison, the goldfish is a much closer cousin to us humans than to the shark. In fact, certain fish with fleshy fins (sarcopterygians), such as the coelacanth, are the origin of tetrapod vertebrates – batrachians, reptiles, mammals – and therefore of humans. Yes, we humans are distant descendants of fish!

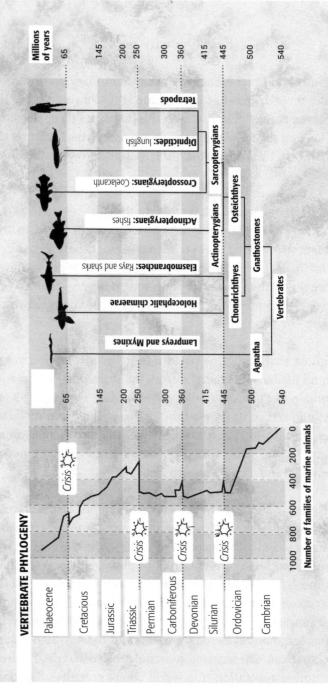

Fig. 6 Geological time and vertebrate phylogeny, where we see that fish are closer to humans than to sharks.

Revolutionary jaws

Are sharks and rays alone on their evolutionary path? Not so simple; there is another companion: the chimaera. Not the phantasmagorical creature with the body of a goat, the head of a lion and the tail of a dragon. No, a real, living creature, the elephant fish (*Callorhinchus milii*), better known as the ghost shark.

In 1987, *Calypso* went looking for it in the sub-Antarctic waters of southern New Zealand. So why are we interested in this chimaera that no one has ever seen alive in its natural habitat? Because it is a prehistoric ghost. Its genome changes so slowly – more slowly than that of any known vertebrate – that it alone can tell us the story of the first ancestors in the chondrichthyan family.[6]

The ghost shark takes us back more than 450 million years, when no life had yet emerged from the primordial ocean and, as Romain Gary said, life did not even dare to imagine itself 'crawl[ing] out of its native mud'.[7] This is the dawn of verte-brates, in a world that is nothing like it is today. Imagine an immense ocean from which only two large islands emerge, 'Gondwana' and 'Euro-America'.[8]

Some marine creatures have already developed an embryonic cartilaginous skeleton. However, the most advanced creatures only had a round orifice capable of sucking, like the hagfish and lampreys, the last representatives of the agnathan group.[9] At this time, a formidable revolution was to transform the world of marine creatures: the invention of the jaw.[10] The gnathos-tomes appeared.[11] Among them were the first chondrichthyans with cartilaginous skeletons, ancestors of today's sharks, rays and chimaeras. These jawed newcomers also acquired paired fins, another revolution in a world where creatures had only one odd median fin along the back and stomach with which to move around.[12]

Palaeontologists, however, have a hard time deciphering the first few paragraphs of this story. For the very primitive chimaeras, which do have the cartilaginous skeleton of chondrichthyans, have smooth skin without denticles and their gill cavities are hidden by an operculum, just like ordinary fish! Evolution is definitely not linear. And life sometimes stutters. It forgets and reinvents, as mutations occur, solutions that it had already proposed millions of years before.

In pursuit of a chimaera

But it's not easy to go back in time. No one has ever come close to a live elephant fish (chimaera), in its native habitat, because it lives on the continental slope, down below 200 metres. Fortunately, once a year, the females return to the coast to lay their strange horny eggs on sandy bottoms accessible to divers. So, for a while, our dream is within reach of our flippers.

8 January 1987. Lat.: 44° 00' S, long.: 172° 30' E. *Calypso* is exploring Canterbury Bay, New Zealand. This morning, the sea is calm, without a ripple. This place is so inhospitable that we are going where divers normally do not dare to venture. From the start, we are gripped by the cold. The water is murky, full of cottony silt, more yellow than dark. In this thick soup, we lose all sense of direction. We swim with our hands skimming the flat, infinitely flat, bottom. Only the craters of a few buried shells and the bulging backs of crabs provide relief on this desolate plain. We feel, rather than see, shapes that slip among us: a fin, the glint of an eye, a spindle-shaped body. Ghosts. Some are as big as my companions, probably blue sharks (*Prionace glauca*) that we have already identified on previous dives. But, intent on our quest, we don't really pay attention to them, nor

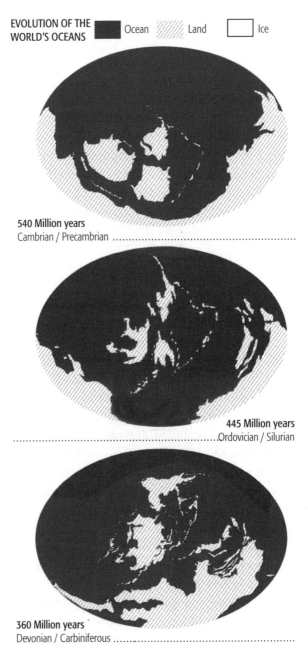

EVOLUTION OF THE ■ Ocean ▨ Land □ Ice
WORLD'S OCEANS

540 Million years
Cambrian / Precambrian ....

445 Million years
....Ordovician / Silurian

360 Million years
Devonian / Carbiniferous ....

Fig. 7 Evolution of the world's oceans.

250 Million years
.... Permian / Triassic

220 Million years
Triassic ....

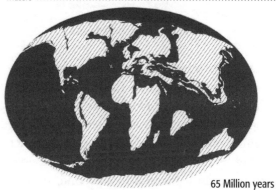

65 Million years
.... Cretacious / Palaeocene

to the huge rays (*Raja nasutus*) that suddenly take off as we pass, raising a storm of mud. One of them disturbs a pod about twenty centimetres long, swollen in the middle. Through the soft brown corneal envelope, I can see an eye. An embryo, a baby chimaera-elephant fish. It is moving! Excitement goes up a notch: the chimaeras are there.

This silhouette swimming along the bottom is not that of an ordinary shark. This fleeting ghost passes us again. Then, slowly, it emerges from the yellow fog. Surprisingly, its appearance is almost comical! The body is indeed that of a fish, but the head seems to have been added on, like a carnival mask from the *commedia dell'arte*. Its slightly conical forehead is prolonged by a disproportionately large cartilaginous nose, which opens out into a large appendage of soft flesh. The astonished look of the large iridescent eyes, the slightly wrinkled skin of the cheeks and the ventral mouth accentuate this mask-like effect, giving the animal the appearance of a sad Pantaloon character. From the front, the large pectoral fins, reminiscent of Babar's ears, transform it into a cartoonish elephant caricature. Our ghost shark may not be more than a metre long. To propel itself, it does not use its caudal fin, but its pectorals, which flap like the 'wings' of an eagle ray.

Hoovering up seafood

The elephant fish (chimaera) ignores us and moves on. Its ploughshare nose seems to sniff the mud. It is not looking for smells, but for the minute bioelectric fields that each living organism generates. Its nose and cheeks are pierced by tiny pores, opening onto receptors sensitive to the micro electric fields emitted by fish, worms or shellfish. And now a clam is paying the price. In a fraction of a second, it is dug up and sucked between the dental plates bonded into the shape of a grinding wheel. This diet was already characteristic of the first

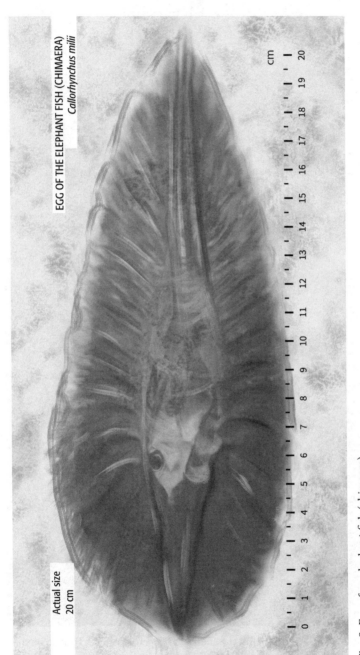

Actual size
20 cm

EGG OF THE ELEPHANT FISH (CHIMAERA)
Callorhynchus milii

cm

0 1 2 3 4 5 6 7 8 9 10 11 12 13 14 15 16 17 18 19 20

Fig. 8 Egg of an elephant fish (chimaera).

Fig. 9 First encounter with the chimaera (elephant fish) off Timaru, on the east coast of New Zealand's South Island, in December 1986.

cartilaginous vertebrates. Thus, the original 'Jaws', the giant megalodons, the terrors of the oceans, had ancestors that terrorized merely worms and shellfish. Armed with microscopic teeth, they sucked or filtered their food.[13]

More likely to be prey than predator, these primitive sharks nevertheless had a few trump cards up their sleeves for survival: the agile swiftness conferred by their cartilaginous skeleton and, above all, a solid spine at the front of each of their dorsal fins. These are formidable weapons that our chimaera has retained and which it uses at the slightest warning. And, all of a sudden, it happens. While gorging itself on a second shell, it seems to discover our presence, and suddenly raises its dorsal spur and flees into the depths of the ocean.

This spur was not just a deterrent; it was deadly. The proof is in the extraordinary 370-million-year-old fossil found in Ohio, USA, which captured the death of a terrible predator: the placoderm *Holdenius* killed by its prey, the primitive shark *Ctenacanthus*, whose spine had perforated its palate.[14]

Cladoselache, the ancestor of 'Jaws'

This act of revenge, inscribed forever in stone, must not have been often repeated. For the first 100 million years of their existence, sharks lived under the threat of predators far more formidable than themselves: placoderms. This group of primitive fish had bodies covered with mineralized plates.[15] Articulated like a knight's armour, they were almost invulnerable. The most famous of them, *Dunkleosteus*, whose stone jaw formed a monstrous shear, was six metres long.[16] Even *Cladoselache*, said to be the most direct ancestor of today's sharks and yet measuring almost two metres, could only flee from such a carapaced creature.

Cladoselache lived in the Devonian, 370 million years ago. It roughly resembled a spiny dogfish shark, hence the commonly repeated idea that present-day sharks are living fossils. In reality, there are no living fossils. The evolution of all living things, however slow, is permanent. Even the elephant fish (chimaera) whose genome evolves so slowly, even the nautilus (*Nautilus* sp.) which resembles the fossils of its ancestors who lived 500 million years ago, are different in physiology, if not in morphology, from all those that preceded them. Often the devil is in the details. For *Cladoselache*, it is the mouth. It opens forward, whereas in modern sharks the mouth opens ventrally. There is also the pectoral fin joint, which is not very mobile and which prevented *Cladoselache* from making rapid changes of direction. Today's sharks use these fins as a pivot to turn on the spot.

Neither *Cladoselache* nor placoderms, not even the super-powerful *Dunkleosteus*, succeeded in getting through the ecological crisis that marked the end of the Devonian period, 358 million years ago, when 75 per cent of marine species were swept away.

A hundred million spectacular years

The field then opened up for a multitude of new species to thrive, giving chondrichthyans 100 million years of fulfilment. Cartilaginous animals had a field day. It would take a whole book[17] to describe all the forms they took to dominate all environments, from the reef ecosystem where the small coral grazer, *Belantsea montana*, was rampant, to the open sea dominated by the *Edestus giganteus* shark, over six metres long. The 320-million-year-old Bear Gulch fossil site in Montana, USA, has yielded more than 113 species, each more extravagant than the last.[18] Behind the head of *Stethacanthus productus*, for example, was a pillar whose top plate was covered with

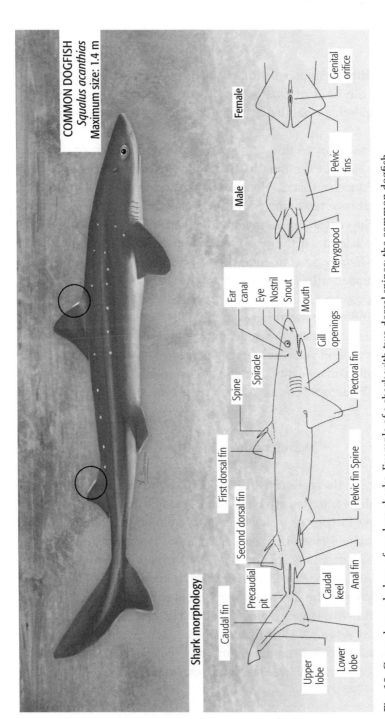

COMMON DOGFISH
Squalus acanthias
Maximum size: 1.4 m

Female

Genital orifice

Male

Pelvic fins

Pterygopod

Ear canal
Eye
Nostril
Snout
Mouth

Spiracle

Spine

Gill openings

Pectoral fin

First dorsal fin

Second dorsal fin

Pelvic fin Spine

Precaudial pit

Caudal fin

Caudal keel

Anal fin

Upper lobe

Lower lobe

Shark morphology

Fig. 10 General morphology of modern sharks. Example of a shark with two dorsal spines: the common dogfish.

denticles. Similar denticles lined the underside of the hook on the head of *Damocles serratus*. Palaeontologists are still wondering what these strange 'toothbrushes' were used for.

Speaking of teeth ... they are no longer implanted around the sides of the mouth. That would be too ordinary for our spectacular sharks. No, the triangular, incredibly sharp teeth must be set in a single row that runs from the gullet outwards in the central axis of the mouth. *Edestus* was thus armed with one scissor on the palate and another on the mandible. As for *Helicoprion*, an eight-metre giant, only its mandible was equipped with a dental spiral, in the shape of a circular saw.[19]

Despite its fearsome saw, you will not find *Helicoprion* haunting our swimming beaches. Like 90 per cent of species, it did not survive the cataclysmic eruptions that, over a period of a million years, submerged present-day Siberia and caused another mass extinction, the largest of all time, at the end of the Palaeozoic era, 251 million years ago.

Sharks in the age of the dinosaurs

But life always returns. The planet, whose geography changed, enabled it to flourish once again. A single ocean, Panthalassa, came to enclose a single landmass, Gondwana and Laurasia combined (see Fig. 7 on pages 38–9). It was on this 'water-world' that most of the species of rays and sharks that we see today appeared. In the Jurassic period, 200 million years ago, while the dinosaurs were establishing their reign on the continents, large hexanchiform sharks, characterized by their six or seven gill slits (instead of five), appeared in the oceans. *Hexanchus griseus*, the bluntnose sixgill shark, their direct descendant, still lives in sea canyons today.

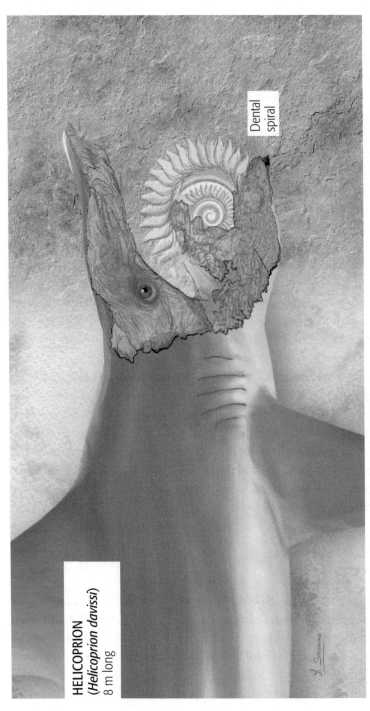

HELICOPRION
(*Helicoprion davissi*)
8 m long

Dental spiral

Fig. 11 *Helicoprion (Helicoprion Davissi)* 8 m long. Artist's drawing based on tomographic analysis of fossil IMNH-37899 by Ramsay J.B., et al.[20]

And it is this 'prehistorical imaginary' that we came to film in the Strait of Messina on 30 July 2011. Why this day? Because the new moon delivers a 'black night' down below underwater. Why here? Because the formidable tidal currents that sweep through the strait between Sicily and Italy cause the water to rise from the bottom of the Ionian Sea in the south, to the Tyrrhenian Basin in the north. The powerful bluntnose sixgill shark hunts in this rising current. The bluntnose should do a better job of taking us back to the Jurassic era than Steven Spielberg! But, as in the film, you don't enter a dinosaur's territory with impunity, especially in the Strait of Messina, between Charybdis and Scylla. In these parts, as Homer sang, 'divine Charybdis sucks black water down./ Three times a day she spurts it up; three times/she glugs it down.'[21] The tidal currents cross a threshold that rises to only eighty metres below the surface in the narrowest part of the strait. Concentrated in these narrows, the flow causes extremely powerful swirls. In ancient times, sailors tried to escape the maelstrom that threatened their skiffs by heading for Calabria, on the other side of the strait, only to end up with Scylla, where the whirlpools are even more violent. Powerful modern ships are no longer in danger, but we divers, like the longboats of yesteryear, remain at the mercy of these formidable eddies. We have barely an hour, before the tide turns, to search the immense strait and find the shark, if it is there.

The traveller in the abyss

Tonight, at a depth of forty metres, 'Jurassic Park' is icy: barely fourteen degrees Celsius, the ideal temperature for the deep water favoured by the bluntnose. In spite of the clarity of the water, the cone of my flashlight seems very feeble in the black immensity of the strait. Like a magic wand, it brings up a thornback skate (*Raja clavata*) and a spotted sea hare (*Aplysia punctata*), and then sends them back to oblivion as soon as it no longer lights them up. Never has the idea that 'only what you pay attention to exists' been so well illustrated. Yet in this Mediterranean, which I know so well, nothing seems familiar. At least, not this strange hydromedusa *Solmissus*, with its thin, deadly tentacles, which the current sticks to my mask. It usually lives in the deepest canyons. Its presence is a reminder that beachcombers regularly pick up lovely hatchet-fish (*Argyropelecus aculeatus*), viperfish (*Chauliodus* sp.) and a whole host of other creatures like it that have risen from the depths. A jellyfish announces the arrival of the shark. But it is too late, the hour is up. The current rises with unprecedented violence, nearly four knots![22] A marine tornado. I am swept away. In a few seconds, the team is dispersed. And by some miracle no one is hurt. The descendants of the Jurassic do not let themselves be seen too easily.

The following night, we are better prepared. Before the under-water storm arrives, we find ourselves face to face with it: a huge square head, framed by large and coarse pectoral fins that are fixed just behind the six gill slits. These six slits and its round back, without the large fin that other sharks have, are the signs of its antiquity. Its very round eye, green like a luminous emerald at the end of the skull, stares at us: an 'alien' facing 'aliens'. Mutual questioning? Except for its size, our bluntnose is exactly like *Notidanoides*, which lived in the Jurassic period, 196 million years ago.[23] Little is known about the bluntnose, which is sometimes seen by exploratory submarines. We do not know where or when it breeds. On the other hand, on this night of the new moon, we are exceptionally lucky to observe a young one barely two metres long, which carries the hopes of the long line of the hexanchiformes.[24]

Prehistoric sharks in the abyss?

How can we explain the fact that the present-day shark has retained a morphology similar to that of its ancestor? The most likely hypothesis is that deep waters are spared the major climatic changes that caused mass extinctions in surface waters. The deep ocean is a stable refuge, which is why other sharks with primitive characteristics (alongside the bluntnose shark) can be found there, such as the frilled shark (*Chlamydoselachus anguineus*), which also has six gill slits, or the megamouth shark (*Megachasma pelagios*), a plankton-eating giant of more than five metres, which was only discovered in 1976.[25]

The depths harbour not just extraordinary sharks. In 1987, off the coast of New Caledonia, located on the ancient continental shelf of the primitive continent of Gondwana, Professor Bertrand Richer de Forges uncovered a field of living pedunculated crinoids (*Gymnocrinus richeri*) at a depth of 470 metres, which were thought to have been extinct for at least sixty-six million years.[26] The nautilus has also survived here, its morphology unchanged for 500 million years. And it is also here, in the Chesterfield Basin, at a depth of 650 metres, that the largest known field of megalodon shark teeth (*Otodus megalodon*) lies.[27]

Megalodon, the tyrannosaurus of the ocean

That was all it took to fire up the cryptozoologists: if there is a shark with a legend on a par with that of the dinosaurs, it is *Megalodon*: sixteen metres long, fifteen tonnes, a mouth in which a man could stand upright, and teeth measuring seventeen centimetres.[28] The perfect monster, the largest marine predator of all time, the *Tyrannosaurus rex* of fish! Like *T. rex* and mammals, *Megalodon* was probably homeothermic, or at least maintained a body temperature well above the

temperature of the ocean. This quasi-homeothermy considerably favours chemical reactions, and thus metabolism. It also
gives the predator speed and endurance, definite advantages
over 'cold-blooded' fish, whose temperature is the same as the
water they are swimming in. Maintaining this body heat is
energy-intensive; it requires a lot of food. *Megalodon* therefore
feasted on whales and even sperm whales, as evidenced by a
fossil found at the Lee Creek Mine site in the USA.[29] This is
probably what caused its demise two and a half million years
ago. When it appeared in the middle Eocene, twenty-three million years ago, *Megalodon* ruled a very warm universe where
it found plenty of prey to satisfy its metabolic requirements.
But the closing of the Panamanian isthmus, separating the
Atlantic and Pacific oceans, caused a major change in oceanic
currents and global ocean cooling.[30] Forced to remain in tropical latitudes, *Megalodon* saw its pool of large prey diminish
considerably. Competition from new, smaller, less demanding
species sounded the death knell for the giant and the advent of
Carcharodon carcharias, the great white shark.[31]

Today's monster

With all due respect to the cryptozoologists, the survival of the
megalodon in the cold depths is unlikely. However, there are
still giant sharks in the heart of the ocean.

I remember a one-metre-high fin which sliced the surface
of the ocean, turning a wave like a plough turns the soil. It was
on 8 July 1981, off the Aran Islands, west of Ireland, on board
the *Petrouchka*, the fishing vessel on which I was collecting
samples for my doctorate in oceanography. This fin was preceded by a sort of whitish bow-spur that regularly pierced the
surface. It was followed at a distance of six metres by another
fin, slightly smaller, which seemed to want to pass it, first to the
right, then to the left. All three were slowly but surely moving

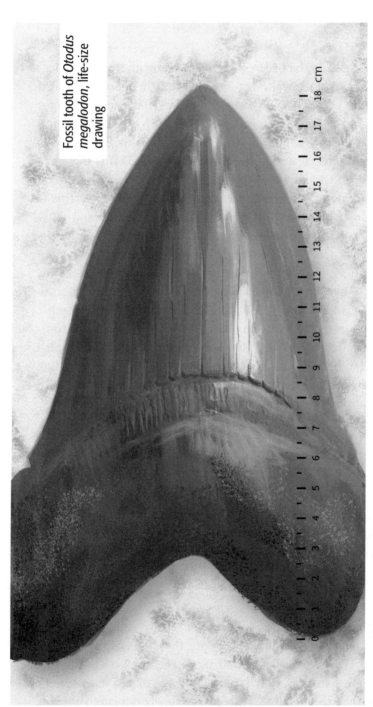

Fossil tooth of *Otodus megalodon*, life-size drawing

0 1 2 3 4 5 6 7 8 9 10 11 12 13 14 15 16 17 18 cm

Fig. 12 Fossil tooth of *Otodus megalodon*, life-size drawing. Wider than the page, it is as large as the dwarf lanternshark; see Fig. 13 on page 55.

towards us. Even the crew of the *Petrouchka*, who had seen it all before, had stopped putting fish on ice. I was fascinated. I didn't understand. Suddenly I saw it through the water: the 'bow' pointing out of the water was just the animal's nose. It was in front of the huge, gaping, whitish mouth that seemed to want to swallow up the sea. The water gushed out on each side through a series of five gill openings. They were so wide open that the head seemed to be partly severed. The massive body followed, inelegantly marked with greyish burrs, as if it had pellagra. On either side was a monstrously thick pectoral fin. Planted halfway down the back, the high dorsal fin. Finally, in the distance, the caudal fin. One and the same animal, a basking shark (*Cetorhinus maximus*). Twelve metres? Thirteen metres? More? I had no idea that such a titanic beast could exist. From the deck of the boat against which it passed, I felt the raw power of this giant. I remembered the poignant images of an incredible hunt for the basking shark, filmed in 1931 by Robert Flaherty for his documentary film *Man of Aran*.[32] A handful of fishermen, in rudimentary boats, struggle for more than two days with the monster, whose resistance is astonishing.

These giants are rare, probably very old individuals whose size can only be surpassed by the whale shark (*Rhincodon typus*), a warm-water cousin that can reach eighteen metres, a hundred times the size of the dwarf lanternshark (*Etmopterus perryi*), which is only eighteen centimetres long (see the life-size shark, pp. 54–5).

In terms of gigantism, stingrays are not to be outdone. In 2015, fishermen caught an oceanic manta ray (*Mobula birostris*) in the Peruvian upwelling,[33] one of the world's richest ocean regions, weighing one tonne and with a wingspan of 8.6 metres.[34] But you don't have to go into the open ocean to see giant rays. The rivers of Southeast Asia are home to

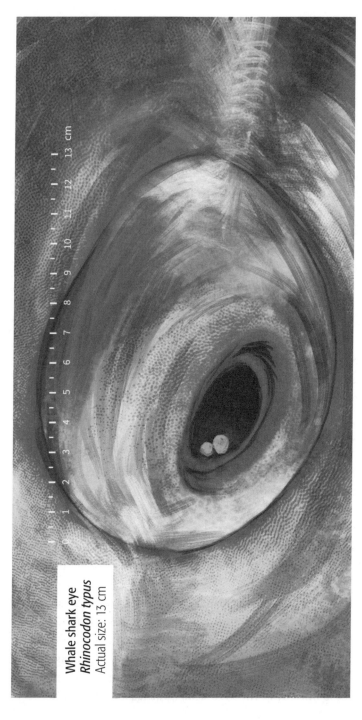

Whale shark eye
Rhinocodon typus
Actual size: 13 cm

Fig. 13 Whale shark eye, actual size: 13 centimetres, and dwarf lanternshark (below), actual size: 18 centimetres.

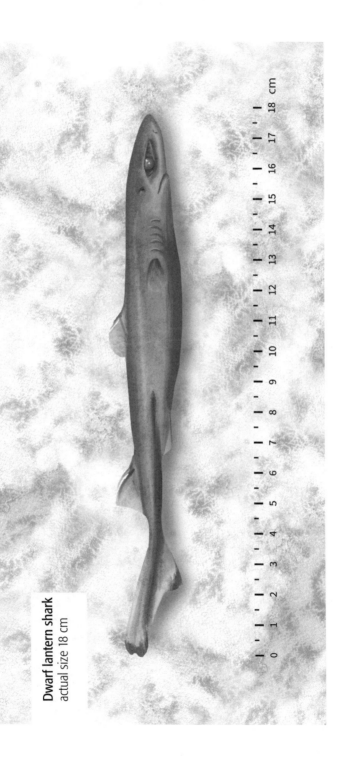

Dwarf lantern shark
actual size 18 cm

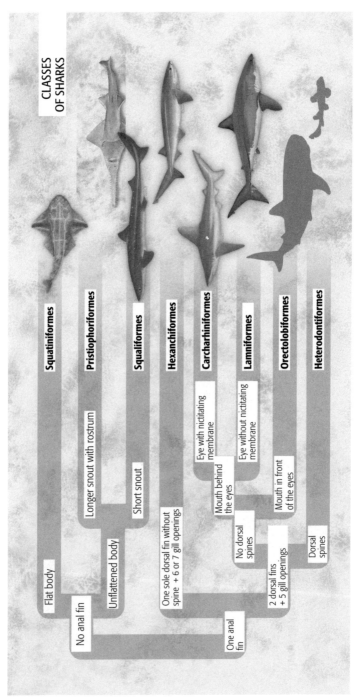

Fig. 14 Key to the identification of the major orders of sharks.

the giant freshwater whipray (*Urogymnus polylepis*), which reaches 600 kilogrammes with a wingspan of two metres.

Natural diversification in full swing

Since the dawn of life 3.8 billion years ago, despite geological cataclysms and mass extinctions, natural diversification has kept individual and specific diversity bubbling away. Chondrichthyans are no exception to the rule and, despite several major extinctions, today number at least 1,199 species, including 536 sharks, 611 rays and 52 chimaeras;[35] other authors estimate 1,125 species, 440 of which are sharks.[36] However, this number is certainly underestimated, as new species are discovered every year, especially since the reformulation of the concept of species based on genetic data has upset the current classification.[37] Notwithstanding the uncertainties about the number of species, there are eight major orders of sharks based on purely morphological criteria that are very easy to observe (fin, number of gill slits). Even the uninformed diver can recognize them. Because you will not fail to encounter them: sharks are everywhere. They colonize all environments, from the Arctic to the Antarctic, from the surface to the abyss, and even rivers and lakes.

I have dived with over forty species of rays and sharks. The diversity of shapes, behaviour and sizes is simply amazing. How can you compare the supple grace of the blue shark, the bluish silkiness of its coat, the sharpness of its round eye whose large iris, black crowned with white, follows your every move, to the nonchalance of an Atlantic nurse shark (*Ginglymostoma cirratum*) whose iris, barely a slit, does not seem to see you?

Is there any resemblance between a ferocious cookie-cutter (*Isistius brasiliensis*), a seventy-centimetre loner, that takes great bites of thick flesh from the sides of giant sperm

whales, and a school of dozens of shy, scalloped hammerheads (*Sphyrna lewini*), whose curved heads swing rhythmically from right to left in search of small squid?

The species are incomparable. There are as many differences among them as between an elephant and a bat! Contrary to popular imagery, no one shark can symbolize all the others.

Fig. 15 School of hammerhead sharks.

3

Giving Life

angrove swamp in southwest Grand Bahama, USA.
January. The dense mangrove forest sends out a few
scouts to attack the seagrass. Their aerial roots are
stuck like stilts in the shallow water. Barely a metre deep. The
water is crystal clear. No wind. The surface is a mill pond and
lets you see the large queen conch (*Strombus gigas*)[1] and the
starfish. In the distance, a discreet eddy progresses over the
seagrass. The dorsal fin of a shark. A lemon shark (*Negaprion
brevirostris*). A huge female with a monstrously distended
belly. She is twelve years old. She has just reached her sexual
maturity. She is primiparous. She swims above the sandy
bottom planted with sparse seagrasses. From time to time she
pauses, then jumps forward a few metres, before collapsing
heavily to the bottom. It looks like she is in agony. Her breath-
ing is rapid. Her mouth gulps water and her gills palpitate
frantically. She is tossed about by the gentle swell. Without any
warning, she suddenly gives birth: her first newborn is expelled
without fuss. Caught between the bottom and its mother, the
baby twists around to free itself. But it remains tied by the
umbilical cord. It takes several seconds for the baby to tear off
the placental cord and disappear into the maze of mangrove

roots. For a 'big predator', sixty centimetres is not very large. The little shark is just 'small fry' among others, at the mercy of its big brothers who prowl around in the evening, or of a barracuda (*Sphyraena barracuda*) that will not hesitate to make a mouthful of a future competitor. The 'forest of the sea' will be its refuge and its larder for the coming year. His brothers and sisters, about ten in number, will join him before their mother, dizzy from the long birth, does not go off hunting for fear of them being devoured.

Very few divers have ever witnessed a calving in natural conditions. The coastal mangroves of the Bahamas are one of the few places known to harbour lemon shark nurseries.[2] Mothers return here that after a twelve-month gestation period to give birth where they themselves were born.[3]

But for other shark species, it is not known where and when mothers give birth. So rare is a sighting, that when surveillance cameras on a remote Philippine seabed accidentally captured the unlikely delivery of the timid thresher shark, the scientific community immediately called for the entire area to be protected.[4]

Yes, the scientific community finds shark births moving, perhaps because they underline our companionship and demonstrate chondrichthyans' surprising originality in the world of fish.

No cuddling

It is not only the flexibility of their skeleton that differentiates sharks from their fish neighbours, but also their reproduction. Like us, sharks mate. Fish, however, have adopted external fertilization. They deliver their sex cells to the currents and the sea becomes their cradle. On the other hand, sharks have opted for internal fertilization . . . But without premarital cuddles or gentle caresses. They are adept at violent and forced penetration.[5] After biting the female firmly near the pectoral fin, the male introduces one of his pterygopods[6] into her cloaca. During the brief embrace, one to two minutes at the most, the joined couple sink slowly. If they reach the bottom before coitus is over, the male will take advantage of this to get a better hold on his terrified mate. Once the coupling is over, she immediately regains her senses and moves away, despite the sometimes bloody wounds inflicted on her by her partner's repeated bites.[7]

Some species, however, seem to be more delicate. With nurse sharks, it is the female that makes advances towards the male. With her pectoral fins firmly planted in the soft bottom, she sweeps the sand from right to left with her genital area, a sort of dance to stimulate her mate or to let him know that she is about to ovulate.[8]

The womb versus the sea as a cradle

We must face the facts: reproduction by internal fertilization with mating, long considered to be very sophisticated and modern because of anthropocentrism, is in fact an extremely ancient type of reproduction among vertebrates. More surprisingly, it has appeared several times in the course of evolution. For example, the first to mate for reproduction were placoderm fish, more than 380 million years ago. Although there have been no witnesses to these unions, which are lost in the mists of time, the fossils speak for themselves. One of them has frozen for eternity a mother of the species *Materpiscis attenboroughi* about to give birth. Her embryos, still connected to her womb by primitive umbilical cords, provided irrefutable proof of internal fertilization, and thus of mating.[9] Other anatomical evidence suggests that the even more primitive placoderm *Microbrachius dicki* must also have mated. The female's cloaca was lined with solid plates on which the male would hook his pterygopods during penetration.[10]

Internal fertilization therefore characterizes very primitive placoderm fish[11] and sharks. External fertilization as adopted by bony fish is therefore derived from internal fertilization and is consequently more modern!

Curiously, everything is in contrast between these two strategies: on the one hand, millions of eggs left at the mercy of the elements, and on the other, a necessarily very limited number of embryos, which, however, are well protected during the critical early phases of development until birth.

Several fathers for one litter

Internal fertilization also limits genetic mixing, which is obviously less rich than when marine currents mix thousands of ova and spermatozoa from dozens of fish reproducing simultaneously. To increase this mix, sharks have adopted multiple paternity: within a single litter, the young can have up to four different fathers. This means that, despite significant injuries inflicted by the male, the female accepts several successive matings.[12] This multiple paternity, which affects all shark species[13] is, however, uncommon, even in species whose females delay fertilization of the eggs by storing the sperm in a nidamental gland,[14] such as the colossal whale shark.[15]

Mermaid purses

The nursehound (*Scyliorhinus stellaris*) is the most patient at this game. It can wait two years before fertilizing and then laying those amazing eggs we see on our dives.

5 August 1972. La Couronne, near Marseille. Forty metres deep. With my friends from the Valence diving club [*Cercle Valence Plongée*], we are looking for a rocky shelf, spotted by depth sounder a few days before. The muddy bottom over which we swim seems limitless, with such poor visibility, without any contrasts, a bit like those foggy days when the sky and the earth merge. Little by little, the horizon darkens: this is the vertical cliff we had hoped for. We are probably the first to discover the life contained in this wall, some fifteen metres high. Into the long cavity that has eaten away its foot, a lobster with monstrous claws and three conger eels point their white-lipped snouts. Under each overhang, the immaculate polyps of the red coral clusters offer a little brightness in the gloom. Lobster antennae bristle on the cliff. There are dozens of them in every crevice. Some of them share their niches with red scorpionfish (*Scorpaena scrofa*), also monstrous, as if overpopulation had forced them to live together. And everywhere, clouds of small pink fish, swallowtail seaperch (*Anthias anthias*), envelop the cliffs in an undulating cloak. This is a Mediterranean that seems implausible to younger generations, so much so that ecological amnesia[16] erases the very idea that nature could have been overabundant.

However, it is not this kind of abundance we find remarkable this particular day. It is the dozens of dogfish eggs hanging in a forest of red gorgonian corals on the top of the cliff. Each one is a horny sheath about ten centimetres long, attractively named 'mermaid's purse', in which the embryo develops. I didn't account for them in my logbook, but I have a clear memory of 'yellow lanterns' in large scarlet fans moving in the current.

Each purse is attached by filaments to the branches of a gorgonian, like a Christmas tree decoration. And the tree is huge:
it's the whole cliff!

Is it a birthing site? I look for the dogfish. They are there,
hidden by the feathery fans. I count fifteen of them, all females.
Some of them, with rounded bellies, are still full. Perhaps they
are waiting for the night to lay their eggs, which will, in turn,
weigh down the branches of the gorgonians and enrich this
incredible Christmas garland.

Most of the purses are covered by bryozoans and calcareous
algae. They were probably deposited several months ago. Some
are shrivelled and empty: the youngster has already gone out to
the open sea. Some of them are still translucent enough to let
you see, against the light, the miniature shark that sporadically
moves around. It is connected by a cord to the 'yolk', the vitel-
line reserve on which it feeds.

You would have had to come back in the following days to
hope to see the hatching, when, too cramped in its envelope,
the little spotted dogfish wriggles in all directions and ends up
cracking the protective envelope. It pushes its snout into the
gap to force its way out; it stops, exhausted by so much effort; it
resumes its fight to push aside the walls of the case. Suddenly,
very quickly, the head, the black almond-shaped eyes, the fins
still soft, and the young dogfish, a miniature and fragile adult,
is wriggling to reach the open water.

A thousand and one ways to reproduce

When it comes to reproduction, nature has invented everything,[17] and sharks have benefited from this. The female zebra
shark (*Stegostoma fasciatum*) can even do without a male!
The egg develops without being fertilized. These cases of

parthenogenesis[18] seem exceptional but have been observed several times in aquaria.[19]

The general rule is that mating is necessary for fertilization. Although 30 per cent of sharks are oviparous, like the dogfish, most have adopted another strategy to protect their few offspring, by preserving the eggs in the uterus throughout their development. The mother does not give birth until the newborn is independent. Here again, there are two possible scenarios: the embryos are fed by the mother thanks to a placenta, which is fairly similar to that of mammals, and this is the placental viviparity of the lemon shark. Another option is for the embryos to develop in the mother's uterus at the expense of a yolk reserve, without any placental link with the mother's body: this is the bluntnose shark aplacental ovoviviparity.

Successful for four million years, but blocked by the Anthropocene?[20]

These different modes of reproduction have thus enabled sharks to survive to the present day. However, the low fecundity that they enable is now proving to be a serious handicap in the Anthropocene. This is especially true since a very long gestation period increases the time required to renew populations: ten months for the bull shark (*Carcharhinus leucas*), which is viviparous and carries a dozen embryos, up to twenty-four months for the dogfish (*Squalus acanthias*), which is ovoviviparous, and even forty-two months for the frilled shark, which holds the record for the longest gestation period in the animal kingdom. This low fecundity is inconsistent with over-harvesting by industrial fishing.

Worse, the rate of extraction by humans is such that it no longer allows sharks to reach reproductive age. The females of most species reach sexual maturity very late: five years for the tiger shark, six years for the bull shark, thirteen years for the sandbar shark (*Carcharhinus plumbeus*), sixteen years for the basking shark. What can we say about the great white shark, whose females are unable to have offspring before they are fifteen, perhaps even twenty-five years old?[21] As for the Greenland shark (*Somniosus microcephalus*), it reaches sexual maturity at around 150 years![22]

Abandoned at birth

This fragility is exacerbated by the immediate abandonment of eggs and newborns, who receive no support from their parents, neither protection nor education. Left to their own devices, they discover and learn about the world at the risk of their lives. This lack of care and cultural transmission makes them more vulnerable than their cetacean competitors, who also have a long life span, late sexual maturity and low fertility, but who take care of their young for a long time after birth. Without the help of attentive parents or a supportive society, how will sharks adapt to the tremendous anthropogenic upheavals facing all living things? This will be the subject of chapter 8, 'Fading Silhouettes'.

A brief digression on 'natural diversity'

It is not unusual for vertebrates to have such a wide variety of methods of reproduction,[23] especially in species that live side by side and occupy almost the same ecological niche. For example, on the reefs of the Great Barrier Reef and New Caledonia, zebra sharks, which are oviparous with the capacity for parthenogenesis, tawny nurse sharks (*Nebrius ferruginus*), which are ovoviviparous, and sharptooth lemon sharks (*Negaprion acutidens*), which are viviparous, share the same sandy coral reef plateaus dotted with marine phanerogam meadows.

Given this diversity, some might be surprised that 'natural selection' has not, over millions of years, retained one good, most effective, mode of reproduction. Is it because they are all equally effective? In the same place, other animals have adopted still more different modes of reproduction that seem to work just as well. Bony fish have sexual reproduction with external fertilization, except for seahorses and syngnaths, which have sexual reproduction with internal fertilization, which places the onus on the male to bear the offspring.

In this large group we also find all kinds of transgender fish: females that turn into males when they get older, or vice versa, or occasionally ... There are fish whose hermaphroditism is obligatory, others not. As for the invertebrates that live alongside these same sharks and fish, their asexual reproduction offers them a major advantage. Some, like corals, play both sides of the fence: sexual reproduction and asexual multiplication by cloning, budding or cuttings. This works so well that these seemingly 'insignificant' beings have changed the geology and history of the planet. If we also look at seagrasses and algae, we can see that living things use a thousand and one ways to procreate, all of which have been effective in allowing the creatures we see today to survive through geological time and mass extinctions.

Natural selection, a positive selection

Does fateful natural selection, which is supposed to keep only the best by eliminating all the losers, as in our culture of competition, not work exactly as we are told?

Unlike artificial human selection, which selects negatively through elimination, natural selection is positive selection. It favours traits that work well and helps to improve reproduction. But it does not eliminate other forms, unless they are really prohibitive and prevent reproduction. This is why we find everything and its opposite in nature. Natural selection passes the living being through the filter of reproduction, and this filter has very large meshes.

Because nature is lazy, it does not select character by character; it operates on a 'morphology–physiology–behaviour' combination. This is the ensemble that is tested through reproduction. If this team reproduces, nature keeps it. And it does not matter if all the components of the team, taken one by one, are not the best. For a highly effective trait can compensate for other traits with dubious benefits. For example, the amazing mimicry of the seahorse compensates for its inability to swim. A human breeder would have done everything to ensure that it was also an efficient swimmer. Nature is unconcerned. The seahorse provides for its succession, that is enough for it!

If natural selection worked according to human practices, we wouldn't be here! Because after 3.8 billion years of selection, there would be only one living being that would have all the right qualities, a kind of Superman. But nature works in exactly the opposite way: it has only multiplied the number of species. With each reproduction there is a small error in copying, a small variant, a small mutation. These micro-differences that selection tolerates even if they are not really useful, nor really any better, gradually make up the diversity of living things, these millions of species that we see today.

Nature entrusts its progress, and its survival on the planet, to the extraordinary diversity of species and their traits.

Perhaps if we looked at natural selection in this way, we humans would think twice about the consequences of our monocultures, which have caused an incredible involution of species diversity. Perhaps we would be more tolerant of other-ness of any kind, and abandon the term natural 'selection' in favour of natural 'diversification', which, after all, defines it much better.[24]

4

Inside the Shark's Head

October 1986. Lawson Bank, Marquesas Islands.

Twenty-five metres deep: the bank is covered with coral debris and dotted here and there with a few large colonies of pocillophores, corals and porites. Giant trevallies (*Caranx ignobilis*) and bluefins (*C. melampygus*), two-spot red snappers (*Lutjanus bohar*) and coral sharks, which have probably never seen divers, spin lively circles around us. Suddenly, these shy carnivores fade away, leaving front of stage to the lord of the place, the silvertip shark. *Carcharhinus albimarginatus* rises from the depths. It is large and bulky: two and a half metres. I am fascinated by its pure body line, its matt sheen, and its calm undulating swim. It has not come alone. Four or five silhouettes appear in the distance and circle around us. Then they melt back into the bluish uniformity and reappear, a few moments later, where we least expect them. The sharks are there, but they are not really there, leaving each of us with an unsettling impression.

How did they know we were there? What tiny vibration alerted them?

Suddenly, one of them charges: it twists, rears up, lowers its pectoral fins and rushes towards us . . . a few metres away, it turns. Such displays are rare. The silvertip shark is usually shy, wary and always on the alert. A sudden movement, a flashlight beam, and it quickly disappears. Dominique Arrieu has to develop a great deal of cunning to get the sharks to approach his camera. The barely perceptible clicking of the film's winding mechanism[1] frightens them. But as soon as the camera stops and silence returns, the sharks, driven by curiosity, reclaim their territory.

We pull up towards an area of coarse sand where blotched fantail rays (*Taeniura meyeni*) are flying low and wide. A strange ballet. These stingrays are over one and a half metres wide and do not flap their wings like mantas or eagle rays. Their large round fins are animated by a long undulation that runs back and forth like a wave that is constantly renewed. The stingrays pass each other, brush against each other, swim together, separate, indifferent to our presence. I wedge myself in the wake of one of them. I almost follow the undulating movement of its wings. I feel the flow of water on my face. It seems to me that I am a fish among fishes . . . Suddenly the great ray stops. It sticks to the bottom like a suction cup and digs its head repeatedly into the sand. I didn't see this coming. I didn't feel anything. I didn't even have time to understand: the stingray had already dug a thirty-centimetre-deep crater and was greedily chewing on its prey. It had sensed the tiny electric field caused by the fish's muscular activity.

This is where we part ways. I will never be a fish, nor a ray, nor a shark, even with a scuba suit. My feeble senses are unable to grasp the environment they perceive. Their universe is much larger than I thought it would be.

Nothing is more different than one ray to another:
shape, swimming, feeding. Here, manta rays filter
planktonic micro-organisms in open water.
Then, a stingray that detects the micro electric fields
emitted by its prey hidden under the sand.

Being a shark

If I were a shark, I would not describe the world as humans do. In the seemingly uniform deep ocean, I, as a shark, would *feel* clouds of odours, electromagnetic masses, hydrodynamic turbulence, that transform the immensity that humans believe to be empty into a liquid thickness charged with information. If I were a shark, I would explain that humans are as inept at describing my oceanic world as a blind and deaf person is at describing the world of those who see and hear. Humans don't even have the vocabulary to describe what I, a shark, perceive.

Like werewolves, half-human half-wolf, who tried to mediate between man and beast, let's try to be 'weresharks' in order to get closer to the feelings of sharks, and enter their heads, their *Umwelt* . . . Let us forget our 'primary senses'. And to do this, let's first close our eyes to be more attentive to the information that our other senses perceive. In addition to the five senses we have developed, sharks have a sixth: electrosensing. But first, keep in mind that it is the synergy of these senses that offers the shark a different world.

Changing the world, changing the frame of reference

Although important on land or in the air, sight is only a very secondary sense in the marine environment. Even off Easter Island, where the water is the purest in the world, free of terrigenous sediments and ultra-poor in plankton,[2] horizontal visibility does not exceed about fifty metres. Usually, in the open ocean, in the sunlit surface area, one can make out shapes at thirty metres, i.e. as bad as on a foggy day. But very quickly, when you get close to the coast in wave-swept areas, near river mouths, in upwellings where plankton blooms thicken the water like soup, the turbidity reduces visibility to a few centimetres. When you go deeper than a hundred metres, night gradually envelops you, never to loosen its embrace. Beyond a thousand metres, there is total permanent darkness, not a photon to be seen, except for the magical lanterns of a few bioluminescent creatures. Ninety-five per cent of the ocean is pitch black. So having the eyes of a lynx or an eagle is no great help. On the other hand, water is denser than air, transports scents better and carries vibrations four and a half times better than air. It is not surprising that sharks have developed, to the point of perfection, organs to sense odours and the slightest variations in water pressure.

The master of smells

The shark kingdom is primarily a world of smells and flavours. The shark is first and foremost a 'nose' that would outperform that of Jean-Baptiste Grenouille, the hero of the novel *Perfume*.[3] Ernest Hemingway, who knew the sea very well, was not mistaken in pointing out that it was the blood of the large swordfish harpooned by the 'old man' that had attracted the first shark:

The shark was not an accident. He had come up from deep down in the water as the dark cloud of blood had settled and dispersed in the mile deep sea. He had come up so fast and absolutely without caution that he broke the surface of the blue water and was in the sun. Then he fell back into the sea and picked up the scent and started swimming on the course the skiff and the fish had taken.[4]

Although no shark can detect a drop of blood in an Olympic swimming pool, as is often claimed,[5] its olfactory abilities are exceptional. It is not so much the concentration of odorous chemicals that matters as the current that carries them to the animal. As soon as the shark perceives the first molecule, it will try to follow the flow of odours to find its source and the meal it promises. It orients itself in much the same way as we humans orient ourselves towards the source of a sound, unconsciously analysing the time difference between the arrival of the sound in the left ear and the right ear. The shark, on the other hand, measures the difference in the concentration of odorous molecules received by each of its nostrils.[6] It always orients itself towards the nostril that is stimulated first. It thus goes from molecule to molecule and always remains in the flow that is densest in molecules, and therefore the most concentrated overall.[7] In this way, it works back to the source. If it goes beyond it, even if it is only one metre, it suddenly finds itself in a smell vacuum and then goes back to find an appetizing scent.

No appetite for human blood

Sharks do not appreciate all odours. They are not very sensitive to human blood, nor to the blood of terrestrial mammals in general,[8] and are particularly attracted to the amino acids in fatty fish (tuna, mackerel, sardines). On the other hand, most sharks seem to fear the necromones[9] released by a dying fellow shark. Numerous experiments have been carried out to use them as a repellent to prevent sharks from biting shipwrecked people, with mixed success. It seems that there are many other factors involved in disasters that involve hundreds of people.

To analyse this vital information, the selachian brain is equipped with impressive olfactory bulbs. They make up almost 50 per cent of the brain. However, not all species have the same perceptive and analytical olfactory capacities. The distribution and number of sensory cells and the morphology of the nostrils, which allow for more or less efficient inflow and outflow, vary from one species to another. Similarly, the olfactory brain areas are more developed in deep-sea hunters, who roam immense spaces to find their food, than in reef species.[10] For example, the great white, which patrols the vast ocean, and the blunt-nose sixgill shark, which searches the immense abyssal plain for corpses, have proportionately larger olfactory bulbs than the coral shark, which lives on a reef teeming with life.

While there is no doubt that, for 'hunters', olfaction is the primary sense, it is not certain that it plays a privileged role in the search for a sexual partner. However, Dr Richard Johnson, who guided *Calypso* on our expedition to the Marquesas Islands in 1986, has no doubt about this olfactory sexual stimulation. During dives he has often observed male sharks with their noses under the female's cloaca, without, however, being able to demonstrate that the male was sensitive to the sexual pheromones of a potential partner.[11]

Taste the water!

In addition to the gustatory sensory cells usually distributed in the oropharyngeal area, there are equivalent chemical receptors throughout the body. The shark thus tastes to its heart's content the effluvia that each creature leaves behind.[12] Before eating, the shark rubs itself against its potential prey, even a whale corpse. It explores with its skin for a long time. It decides on the basis of the flavour it detects. It will not bite if the taste is unpleasant or unusual. This is why some particularly daring sharks sometimes come to rub up against divers. This is of no consequence, especially if the diver remains calm. The shark will turn away because of the unfamiliar taste of the wetsuit.

Hearing with the whole body

No pinna. The ear is just a tiny orifice barely visible behind the eye. And since there is no eardrum, the inner ear is mainly used for balance. The shark hears with its entire body, thanks to thousands of hair cells housed in a canal, called the lateral line,[13] which runs along each side from the head to the tip of the tail. This channel, filled with mucus, is open to the outside environment through hundreds of pores. The slightest wave caused by movement or noise excites the cells sensitive to minute variations in pressure, immediately putting the shark on the alert.

In addition to the lateral line, there are other hydrodynamic mechanoreceptors, spread out over the rest of the body, which measure the flow of water over the skin. They give the shark a sense of speed and orientation relative to the current. These receptors detect fluid displacement speeds of the order of a micrometre per second.[14] They are also very sensitive to low-frequency vibrations, below 200 hertz, preferably around 40 hertz, especially if they are discontinuous pulses. These low-frequency waves, which are inaudible to us, travel very far and very fast in water – 4.5 times faster than in air. They travel evenly in all directions, unlike scent molecules, whose dilution depends on the current. The shark can thus be alerted by the

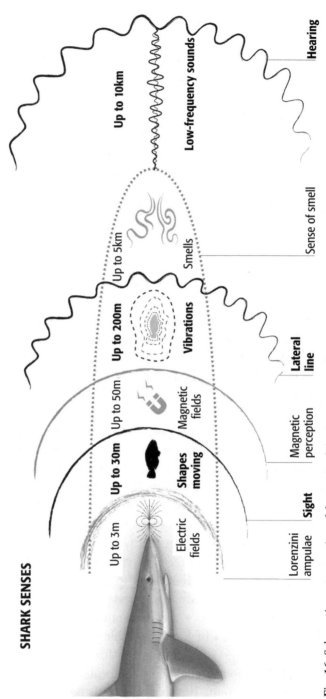

SHARK SENSES

Up to 10km — Low-frequency sounds — **Hearing**

Up to 5km — Smells — Sense of smell

Up to 200m — **Vibrations** — **Lateral line**

Up to 50m — Magnetic fields — Magnetic perception

Up to 30m — **Shapes moving** — **Sight**

Up to 3m — Electric fields — Lorenzini ampulae

Fig. 16 Schematic representation of the range of shark senses.

chaotic swimming of a wounded fish, even if it is upstream of the fish, whereas it cannot smell the odour carried downstream by the current. Better still, this lateral line is paradoxically necessary for the shark to find a scent source that does not emit any vibrations! If the shark loses the first scent, it can simply swim upstream using its pressure sensors to stay in the heart of the current and find the prey with the enticing smell. Experiments with the dusky smoothhound (*Mustelus canis*) have shown that, without their lateral line, sharks have very little chance of finding the source of the scent.[15]

In contrast, sharks are not very sensitive to continuous sound or high frequencies.[16]

Thanks to all these pressure sensors, the shark has a permanent and complete dynamic map of all the fish moving within a hundred to two hundred metres of it, as detailed as the one we have with our eyes.

Seeing movements better than forms

Sharks are not 'visual'. But when you feel the gaze of a bull shark or oceanic whitetip shark (*Carcharhinus longimanus*) settle on you, you know it will never let you go. Like an anthracite ball rolling in its socket, it follows you as if magnetized . . . At close range, sight is an essential asset, especially as the shark's field of vision, with eyes on either side of its head, covers 360 degrees whereas ours is barely 200 degrees. This advantage, however, requires monocular vision, which is less precise for assessing distance than our binocular vision. This is all the more true because the eye's accommodation system is not based, as in our case, on changing the convexity of the lens, but on the movement of the lens from front to back in relation to the retina. It seems that most species only see well at medium distances, not near or far.[17]

On the other hand, sharks can see in the dark, thanks to a layer of mirror cells, called *tapetum lucidum*, which lines the back of the eye and reflects even the smallest photon to stimulate the retinal cells. These reflective cells can be masked by black pigments, so that the shark is not dazzled when it is close to the sunlit surface. This device allows the shark to adapt fairly quickly to light variations and to compensate for the extremely slow retraction/dilation of the pupil. Here again, there is great inequality among species. For example, deep-sea species have enormous difficulty adapting to changes in light intensity.

The bluntnose sixgill shark that we filmed at night in the Strait of Messina, totally blinded by our lights, seemed unable to react to the glare. Conversely, I have often seen great white sharks poke their heads out of the water seemingly to observe us, suggesting that their eyes can cope with the great difference in light levels on either side of the surface. However, there is no evidence that they are able to distinguish shapes in the air correctly.

Life in black and white

As with other vertebrates, the retina of sharks is made up of cones and rods, cells which, theoretically, should allow colour vision. This is believed to be true for skates.[18] Sharks, on the other hand, seem to have a monochromatic vision[19] that varies with the proportion of different retinal pigments: rhodopsin (blue/green sensitive) or porphyropsin (red-sensitive). Deep-sea sharks also have a deep-blue-sensitive pigment, chrysopsin.

Finally, the eyes of sharks of the order Carcharhiniformes[20] are protected from shocks by an 'eyelid', or nictitating membrane, which covers the cornea from bottom to top. It is this whitish eyelid that the shark closes just before biting, giving the unpleasant and confusing impression that it is blind.

Sixth sense

All chondrichthyans have an extraordinary sixth sense that we lack, electrosensing: the ability to detect extremely weak electromagnetic fields. These electromagnetic fields have two origins: terrestrial magnetism, which sharks use to orient themselves during their migration (see chapter 7), and the electrical potential differences linked to the activity of each living being. For example, the electrical activity of neurons, which doctors use electroencephalograms to detect, and the electrical activity of heart muscle cells, which they use electrocardiograms to detect. Rays and sharks, in the same way, can easily detect an immobile organism buried in the sand, invisible to ordinary predators and humans.

These minute electromagnetic fields are detected by organs called 'ampullae of Lorenzini'.[21] They are distributed all over the shark's head. Their walls are made up of sensory cells that are immersed in a jelly. They evaluate the difference in electrical potential between the shark's internal environment and the local terrestrial electromagnetic field that serves as a reference. The shark thus instantly perceives the disturbance caused by the bioelectric activity of a living organism. The best at this game is without doubt our elephant chimaera. It has developed its arsenal of Lorenzini ampullae at the expense of odour receptors.[22] It is past master at detecting worms, shellfish and other small molluscs buried in muddy areas.

All for one, one for all

The senses work together. The keenness of each of the sensory systems, taken independently, does not really account for the shark's real perceptive capacity. The synergy of all the senses is far greater than the sum of their individual capacities. The

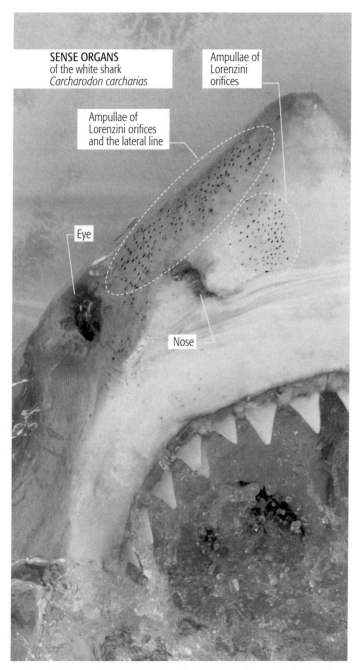

SENSE ORGANS
of the white shark
Carcharodon carcharias

Ampullae of
Lorenzini
orifices

Ampullae of
Lorenzini orifices
and the lateral line

Eye

Nose

Fig. 17 Pores of the ampullae of Lorenzini distributed over the head of a great white shark.

shark uses them both in relay and simultaneously. Sight, and electrosensing as a final resort, confirm the initial location of the prey, identified by its smell and the low-frequency waves due to its swimming movements and the batting of its gills. It is often difficult to know which sense is being used in the action, as their complementarity allows the shark to compensate for the failure of one by the other. Even as it bites and swallows its prey, when touch and electrosensing have taken precedence over the other senses,[23] this complementarity is still necessary: the prey will only be swallowed if it smells good and is appetizing.

Although each species modulates the use of its senses, the sense of smell remains *the* primary one. However, laboratory experiments show that deep-sea hunters and shellfish-eaters do not favour the same sequence. The blacktip shark (*Carcharhinus limbatus*), which hunts pelagic fish, favours the sense of smell, then vision, before the lateral line takes over. The bonnethead shark (*Sphyrna tiburo*), which detects crustaceans olfactorily in the seagrass beds, only uses vision, then electrosensing, when biting and swallowing. The nurse shark also uses its sense of smell to find fish buried in the sediment. But vision, lateral line and finally touch follow one another to refine the pursuit of the prey, until ingestion, which brings in electrosensing.

Like a second skin

This arsenal of perceptive faculties not only serves the 'predatory shark', it also serves the 'social shark' and the 'prey shark'. For it defines a kind of 'tactile' envelope around the animal's body, inside of which a stranger cannot penetrate without triggering a lively reaction, as if the shark were physically touched. This aquatic sensory bubble, which can be described as a

The bull shark's sphere of intimacy. The shark maintains a safe distance from the divers Steven Surina and the author, although they are perfectly still.

Fig. 18 Privacy sphere of a great white shark. Here the author is at the edge of the alert zone. The shark will change direction to maintain its safe distance.

'sphere of intimacy', determines a comfort zone and an alert zone, two to three metres from the physical body. The radius of this 'sensory shield' varies according to the species, the time and even the individual. In general, outside of feeding frenzies, sharks do not touch or stroke each other (unlike marine mammals) or even brush against each other. Two sharks that pass each other move apart to maintain a comfortable distance. The same applies to divers. And you have to develop a great deal of calm and consideration if you want to swim shoulder to shoulder with a shark, as I have had the chance to do several times with great white sharks, as they only allow you to enter their alert zone on exceptional occasions.

In the head of the shark

What do sharks think about fish? What characterizes a human being, for a shark? Its heartbeat more than its shape? The appetizing micro-discharges of electricity it emits? The disorderly way it swims, so different from other sea creatures? The unpleasant taste that permeates the water as it passes?

How can we put ourselves in the head of the shark, when we can feel neither the earth's magnetism, nor its electric fields, nor the variations in ambient pressure? How can we describe its perceptive universe, when we don't even have the vocabulary to express it? We find ourselves in the situation of a man born blind who has to describe the world of the sighted.

Worse, even if we were to identify perfectly the information received by the shark, we would still be unable to account for the workings of its mind. Its 'phenomenal consciousness'[24] cannot be reduced to the physical or physiological conditions of its appearance. For the physical and physiological feeling is compared to the reference frame acquired in the course of one's own history, which, moreover, is permanently modified by one's experiences. Phenomenal consciousness is inseparable from the individual environment. It cannot be objectively grasped by physiological measurements.

In fact, we can only imagine the perceptual universe of the shark in reference to our own perceptual universe.

So how do we understand the shark? This is the question posed by the philosopher Thomas Nagel, in his wonderful article 'What Is It Like to Be a Bat?'[25] Here Nagel develops the idea that we have absolutely no way of knowing what experience of the world a bat has; the only way would be to be a bat yourself.

We will never be sharks. But are we then condemned to remain locked in irreducibly compartmentalized worlds? Can we not hope to find 'points in common', communicative bridges, in order to live together? This will be the subject of chapter 10, 'Reconciliation'.

5

On the Road to Personality

Caribbean Sea, Playa del Carmen. 5 December 2019, twenty-four metres deep. My gaze is hypnotized by dunes rolling on endlessly. They are miniature dunes a few centimetres high, whose crests, wafted by a southern current, melt like watercolours into the watery blue of the depths. Vertigo. For half an hour, I have been sitting motionless on the sandy bottom alongside Steven Surina, a specialist in human–shark interaction. I scan the empty landscape. Nothing. Nothing to break the monotony.

And suddenly, we are sure of it, without being able to pick them out: they are there. Undefinable but present. They are all females in Playa del Carmen, bull sharks, massive, paunchy, not really attractive at first sight. The mere mention of their name should frighten us, as it conjures up the most terrible fantasies of man-eating sharks. But it's the opposite; we are relieved, as we have been waiting for them for six days and twice as many unsuccessful dives.

In the distance, a form takes shape. Two and a half metres, 150 kilos: it's Ana. She swims slowly, powerfully, confidently. She does not change direction. Without warning she cuts off Stella, another adult female who turns ninety degrees and

Fig. 19 Ana, a female bull shark, swims with determination over the sandy bottom of Playa del Carmen. She is characterized by her dark coat, extending well below where her pectoral fin joins.

Morphological characteristics allowing the identification of some of the bull sharks of Playa del Carmen.

fades into the blue. Ana and Stella are the same size. Yet the former dominates the latter. There is no doubt about the difference in character between the two adult females. There is no doubt about the difference in boldness: while Stella seems to avoid us, Ana comes straight at us. Steven and I have been dreaming of this moment for a long time. We press ourselves against the bottom to appear less conspicuous. We don't want to impress Ana; we don't want to enter her 'sphere of intimacy' so she doesn't change her route at the last moment. We hold our breath.

I even avoid looking at her directly. Our discretion works; Ana's senses are not alerted. Her curiosity overcomes her defence mechanisms. She comes closer. We try to become even more inconspicuous. Ana passes right over me. It is a great moment.

I look at her dark coat, which extends well below where the pectoral fin is attached. I notice the cut on her left pelvic fin. These two features distinguish Ana from the other females. Another, Ayaté, has a very wide and curved dorsal fin that cannot be mistaken for Wave's, which has lost its tip. In Martine's[1] case, it is the lower tip of her caudal fin that is cut off in the shape of a sickle. These physical markers make it easier to recognize the forty or so individuals that frequent the waters of Playa del Carmen every year between November and March.[2]

Steven, who has been diving with them since 2013, can even tell them apart by their behaviour, their position in the water, their attitudes towards their sister sharks. 'Each one has its own personality', he points out, with a twinkle in his eye. Steven is happy. He wants so much to show the world that these shy bull sharks deserve better than blind human hatred. He wants to show that you can not only be around them, but understand their codes. He wants to be a shark among sharks, to get inside their heads, to understand the world like they do.

'Shark "personalities"?! That's fine with monkeys, but with elasmobranchs, you must be joking!' retort the behaviourist biologists who claim that rays, sharks and other vertebrates, described as primitive,[3] are beings without identity who react in standard ways to variations in environmental stimuli.

Is Steven so hypnotized by the sharks that he has lost his sense of judgement? Having dived with hundreds of rays and sharks, I have, like Steven, an intuition that each one has its own identity, perhaps even a form of personality, as Eric Clua said back in 2010, when he highlighted very significant differences in predatory behaviour within a population of great white sharks.[4]

But what are the objective elements that reveal these personalities, and how do we define them?

A stingray in love

Every diver has, on some occasion, experienced an improbable encounter that undermined his or her certainties. For me, it was a manta ray (*Mobula alfredi*) on 15 March 1990 at Stort Reef, in the Mentawai Sea off Sumatra. A few hours later I wrote in my logbook:

> This morning I was alone in the water. The manta arrived in front of me, perfectly presentable in her black dress with a white front. This one-tonne giant had the superb grace of an albatross, the precision of a swallow and the majesty of a golden eagle. She glided through the water effortlessly. The water that resisted me seemed to propel her. I froze so as not to frighten her. For, if the slightest thing can intrigue her, a small movement can make her flee. And more than anything else, she fears physical contact. Any yet, she came closer and started to dance. She twirled on her back and then on her side. She swept in ever tighter arabesques, as if she wanted to hypnotize me. She brushed against me with her wing and offered herself for my caress. I was afraid to break the spell, but I caressed her soft, rough skin, like fine sandpaper. She swam backwards in a wide loop, then circled out to the side, coming back to face me. I stroked her again and she took off again in a waltz. I caressed her again and again. When I ran out of air and had to come back up, she was the one to come looking for me. She got in my way, tried to hold me in her wings as if in a cape. Two flips by way of farewell, and she vanished into the blue.

For humans, cold-blooded animals like reptiles, amphibians and fish are just good for killing, laying eggs or being eaten.

What do we know about their behaviour? We think they are stereotyped, bent on a single objective, surviving in order to reproduce. We want to see every action as an innate exercise in getting stronger. Yet certain behaviours sometimes seem to have no immediate objectives, just a waste of time and energy.[5]

Cold-blooded animals, standardized robots?

Since then I have often observed our warm-blooded cousins, dolphins and sperm whales, in their frivolous games, their pointless chases to grab an alga or a tree trunk,[6] so that I cannot avoid the question: what is the threshold of complexity that authorizes playful behaviour, useless behaviour with no direct link to survival? Where is the boundary that pushes so-called 'lesser' living organisms into the darkness of standardization? Is it not simply a lack of attention that has led us to believe that cold-blooded vertebrates, with their standardized reactions, are all alike? Have we ever looked for their identity, their capacity to innovate, or to build a personality?

The ethologist Gordon Burghardt describes play as 'any behavior not necessary for survival, performed for its own sake, without stress, and which consumes time and energy at the expense of survival activities'.[7] Burghardt sees play as the motor of individual construction and innovation. He recognizes this ability not only in mammals and birds but also in amphibians. So what about sharks and rays?

Neuroscience is making steady advances in showing not only that, in the animal kingdom, cognitive capacities are abundant, but that sensitivities, even emotional ones, are a continuum within the living world. Moreover, elasmobranchs seem to have all the cerebral capacities to 'escape' from the strict confines of necessity.[8] For example, oceanic manta rays (*Mobula birostris*) have the largest brains of all elasmobranchs and fishes combined. As for the *Mobula japanica* ray, its telencephalon[9] accounts for 61 per cent of its total brain mass.[10]

Cooperative fishing

These colossal rays, which can have a wingspan of more than seven metres[11] and live for decades, feed on tiny planktonic creatures that they filter through their gills armed with branchiospines[12] that function like the teeth of a comb. To optimize the capture of these tiny organisms dispersed in the liquid thickness, the manta ray deploys two gutter-like cephalic appendages in front of its mouth and twists about to create a vortex that traps the plankton. More often, however, it is a giant vortex that several rays create by swimming in single file above each other. The gaping mouths follow one another like an armada of trawlers, leaving no hope for the most tenuous of brimborion. They seem to cooperate so that

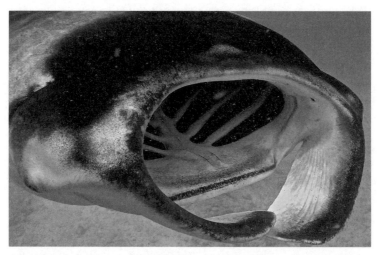

Fig. 20 The gaping mouth of the manta ray. The cephalic horns (the two pallets on the front) form a funnel to channel the microscopic plankton (small white dots) towards the gill slits.

each one feeds as well as possible with the least amount of energy.

In August 2017, in Hanifaru Bay (Baa Atoll, Maldives), with my wife Véronique, we witnessed a Dantean feast of mantas for two hours. There were ten, twenty, fifty of these giant rays. They appeared one after the other in the greasy green water, over-loaded with plankton. In single file, slightly offset one above the other, they formed a sort of giant creature with multiple mouths that swallowed everything in its path. An endless merry-go-round of disproportionately gaping mouths, unabashedly exposing the pearly velvet of their split throats on both sides of the five gill slits. They brushed against us, tilting their large wings at the last moment to avoid hitting us. Thoughtful. The black backs followed the immaculate white bellies and seemed to roll over themselves to ride the current and come back again and again. A hypnotic ballet. Everything seemed orderly. The rays swam together, at the right pace, at the right distance, to optimize the flow of plankton to the benefit of all.

Two hours was not enough time to identify each ray, to dis-tinguish among leaders, followers and beaters, to support our hypotheses on the organization of the group. In contrast, Robert Perryman's team spent five years identifying and monitoring more than 500 groups of manta rays in the waters of Papua New Guinea, in the privacy of their cleaning stations as well as at the heart of their orgiastic feasts. The team highlighted the social links that structure these gatherings. It revealed the elective affinities that unite this and that manta, demonstrating the strong identities that characterize each one.[13]

The dawn of a social network

In sharks, the differences in 'character' among species are very marked. Divers experience this every time they go down.

You have to be extremely discreet not to frighten a school of scalloped hammerheads, whereas, as J.-Y. Cousteau liked to tell us on board *Calypso*, the oceanic shark (*Carcharhinus longimanus*) is 'the only member of their species that is never really afraid of divers ... the only one whose inquisitive and incredibly self-confident temperament led Didi [Frédéric Dumas] and me to get out of the water'.[14] But these differences also exist between individuals of the same species, within the same population. As we have seen, each bull shark in the tiny Playa del Carmen population has its own character traits: Stella does not behave like Ana, Martine, Wave or Julia ... And what is true for the big sharks is also true for the smaller species. Evan Byrnes has shown that each small Port Jackson shark (*Heterodontus portusjacksoni*) reacts differently to the unknown. Some are very shy, while others systematically and boldly explore a new environment.[15] The same is true of lemon sharks (*Negaprion brevirostris*): there are the shy, the bold, the curious, the explorers, the placid and the hyperactive. Some like to live in groups, others are more aggressive towards their fellow creatures. Each individual shows a combination of characteristics that is truly distinctive.[16]

These differences in character create a hierarchy within the school of sharks that constitutes a basic social network, known as 'fission–fusion'. Some sharks choose who they want to be with and play a leading role in maintaining 'social' relationships within the group.[17] The bonds are much less complex and lasting than in cetaceans, such as sperm whales.[18] However, some individuals maintain close, long-term privileged relationships. Johann Mourier has shown that in the population of blacktip reef sharks (*Carcharhinus melanopterus*) in the lagoon on the island of Moorea, some individuals, who are not related to each other, have very strong affinities and do not have any particular relationship with others, even though they are in contact with them every day.[19] These results raise a host of questions that no one has been able to answer to date: do sharks have forms of attachment to others? Do they have emotions? Do they have empathy? Do they experience stress when separated from those with whom they bond? In any case, these results can only be understood if we accept the idea that, beyond its genetic identity, each shark has its own personality.

Identity, personality

It is important we understand what is meant by 'personality'. Every shark, like every living being, has its own genetic identity, i.e. its own perceptive and analytical potential, etc. And it is on this static basis that personality is developed. Personality is a *dynamic* construction that is constantly changing and is based on the permanent dialectical enrichment of its *Umwelt*: the being lives an experience to which it responds with its current experience. This experience changes it and this change in turn impacts its environment, etc. Twins have the same genetic identity, but irrevocably build different personalities according to their personal experiences. Of course these experiences are more or less rich.

Abandoned at birth by their mothers, sharks do not benefit from their parental education like cetaceans. They are not very social and are not very stimulated by their fellow creatures. Fortunately, most species have a long lifespan, giving them all time to gain experience. But one needs to have a good memory to learn from one's experiences . . .

Do sharks learn? There is little research to answer this question. Vera Schluessel's team has shown that captive grey bamboo sharks (*Chiloscyllium griseum*) are able to recognize shapes in order to obtain food as a reward. However, individual learning abilities are extremely uneven: some individuals remember shapes for more than fifty weeks, while others never complete the exercise.[20] Johann Mourier also notes that blacktip sharks that have been caught once are more difficult to recapture. In other words, they have memorized their misadventure and learned from the painful experience.[21]

This learning process is particularly crucial when the juvenile shark changes its diet as an adult. For example, when a young white shark, which had previously consumed only small fish, reaches three metres in length, it adopts a much more varied diet, which includes other sharks and pinnipeds.[22] To do this, each individual tests and adopts its own hunting strategy, independently of its cohort.[23] During this short period, when its exploratory abilities are particularly put to the test, the young white shark tests all kinds of prey. Eric Clua believes that if the young shark is confronted with humans during this dietary change phase, it may not only remember them but also consider them potential prey, which will facilitate a possible bite attempt on a human.[24] However, this exploration phase diminishes with age, considerably limiting learning and thus personality construction.

Environmental complexity and the expression of personality

In fact, the more we study sharks, the more we can highlight those individual variations that do not fit the crude descriptions that have prevailed until now.[25] Kátya Gisela Abrantes, for example, shows that nothing resembles a broadnose sevengill shark (*Notorynchus cepedianus*) less than another broadnose shark, as individuals seem to do what they like. Within the same population, each individual seems to act and move about as it pleases and, above all, decides on its own diet.[26]

Better still, the more complex and dynamic the ecosystem, and the more it is subject to change, both natural and anthropogenic, the more individual variability is expressed, and the more personalities are intensified.[27] The lack of food, caused by human overexploitation, pushes sharks to seek out new prey, even very unusual ones. The most reckless and exploratory sharks, who try to vary their food resources any way they can, accentuate their differences with other individuals. The emergence of new food and ecological preferences can, in the long run, cause a split within the species and even be a source of speciation.[28]

Play as an evolutionary driver

The explorations that build the personality of sharks have a limit, however: they seem to be motivated solely by questions of survival, the search for food or hiding places to protect themselves from predators. Strictly 'utilitarian', they are less stimulating than the free exploration offered by 'useless'[29] play, which opens up all possibilities. To date, no one has demonstrated that sharks explore for pleasure, out of simple curiosity. No one has observed sharks playing. Nor is it known how sharks spend their 'free time', the time they do not spend on activities strictly necessary for survival. Perhaps this is why, despite their great individual diversity and strong identities, lemon sharks show very little plasticity and very little capacity for improvisation.[30] Indeed, just because the shark tries to increase the range of its diet does not mean that it is very curious.

True curiosity, such as that shown by cetaceans, has no immediate goal. During our eight years of underwater observations of the same sperm whale clan on the west coast of Mauritius, we have repeatedly observed young and adult sperm whales playing with floating objects, coming to explore our boat or, better still, coming to meet us and stay with us, just to satisfy their curiosity. These stress-free explorations, which do not serve a survival imperative, can last for hours, whether alone or in groups. They are done without any constraints or immediate objectives. What the sperm whales learn from these free moments may never be used, but sometimes, much later, in circumstances that the animal had obviously never considered, they can prove to be a source of inspiration, a solution to a problem or a source of innovation. The story of Eliot the sperm whale, who insisted that divers remove the hook that was hurting him, is a good example. Eliot had had to learn to meet people in order to satisfy his curiosity so that years later he would return to those he knew to ask for help.[31] Arthur, his fellow whale, who had the same misadventure but was much more shy and always stayed away from divers, kept a hook in his mouth for days without ever asking for help from humans. Two different personalities constructing themselves via their personal experiences.

Self-awareness, awareness of others

Is the shark aware of itself and, consequently, conscious of alterity? To answer this question, Csilla Ari conducted the famous mirror test on two manta rays.[32] Without success. Nevertheless, Csilla Ari considers that the unusual attention that these rays paid to their image would be an indication of the beginning of self-awareness.[33] It may be surprising to see this test, designed for animals whose primary sense is sight, still being offered to species for which sight is a secondary sense. They therefore have no reason to pay attention to an image. It's a bit like giving the mirror test to a blind person . . . what conclusion can you draw?

As we do not have a test that can demonstrate this consciousness, we will adopt here the definition of animal consciousness proposed by Pierre Le Neindre in the report of the European Food Safety Authority: 'a biological phenomenon, that of the reflexivity of the individual on itself'.[34] When we say, the shark feels pain,[35] this implies that it has, at least, this reflexivity about itself and, therefore, awareness of itself and the external environment.

What can be said about this capacity to feel oneself as a being distinct from one's environment, and which most soberly defines self-awareness, is that this awareness is not monolithic. It is not an 'all-or-nothing' phenomenon, but a progressive emergence, which allows one to distinguish oneself from the environment until one is aware of what others are aware of (theory of mind). Neurobiologists prefer to speak of a cluster of skills that shape this awareness. This consciousness is expressed more or less intensely because it relies on more or less complex cognitive systems and on equally varied analytical capacities for 'received images'. We can even say: 'there are as many individuals as there are consciousnesses'.

Yet there is a vague sense that the shark, like the snail, does not have the same capacity for consciousness as a human. But to date, we are not in a position to assess this difference. For the absence of proof is by no means proof of absence of consciousness. Perhaps we do not (yet) have the means to measure the consciousness of sharks, snails or insects. Perhaps we are not looking in the right direction, nor asking the right questions to reveal a consciousness that is too subtle and too different from our own. Let us not forget that until 1987, the human infant was considered to feel neither pain nor suffering during its first months, because doctors were simply unable to measure them.

Understanding each other's thinking?

To date, no experiment has shown that sharks have the ability to understand each other's thinking in order to join forces and take joint action.

However, the naturalist Russell J. Coles recounts observing bull sharks (*Carcharhinus leucas*) off Cape Lookout in North Carolina chasing a school of bluefish (*Pomatomus saltatrix*) in a highly coordinated fashion, herding them towards the beach. This hunt requires anticipating the escape of the fish into the water column and therefore knowing how the other sharks will act, so that the fish, which are much faster than the sharks, cannot escape. This hunt is therefore much more complex than the cooperative fishing of manta rays, which must simply position themselves in relation to each other, without having to anticipate the movements of the other rays.

More exceptionally, on the coast of Argentina, divers have observed sharpnose sevengill sharks (*Heptranchias perlo*) hunting in groups for dusky dolphins (*Lagenorhynchus obscurus*) that are much faster and more powerful than themselves. This hunt involves cooperation, consultation and anticipation of each other's movements in the absence of communication by sound (sharks do not make sounds). Do these observations suggest that, as with dolphins and killer whales, these sharks are capable of representing another mind to themselves?

No one has detected a hint of empathy in sharks either, the ability to put oneself in another's shoes, to suffer when the other suffers, to feel joy when the other is happy. Moreover, the so-called mirror neurons that are responsible for this empathic capacity have not been identified. No signs of mutual aid, altruism or particular interest in members of one's own species or even group have been detected.

The shark seems even less capable of meta-cognition, that ability to know what we know, to be aware of our knowledge and especially of its limits. Here again, we cannot help but compare it to the other kings of the ocean, the cetaceans. The bottlenose dolphin (*Tursiops truncatus*), for example, is aware of what it can and cannot do. Subjected to an increasingly complex test that leads to an impossible choice, the dolphin refuses to choose and, when given the opportunity, opts for a neutral choice, demonstrating that it is aware of the limit of its discernment abilities.[36]

Passing on experience

Culture does not seem to be part of the elasmobranch world either. The development of a culture is all the more improbable as there is no intergenerational transmission, no education of young by their parents, and relations between individuals are quite unstable. However, there could be unintentional transmission of knowledge acquired by imitation, even in non-social animals. The experiment was carried out by Tristan Guttridge's team on lemon sharks[37] and Catarina Vila Pouca's team on Port Jackson sleeper sharks, which are solitary in their early years but may occasionally be found in groups.[38] Vila Pouca put together naive individuals and those who had been taught a route to find food. The ignorant ones all benefited from the knowledge of their learned counterparts by imitating them. They were infinitely more successful than control sharks who had no role models to emulate. But the individual disparity led to strong differences in learning; some sharks learn and change their behaviour permanently after memorizing the knowledge of experienced fellow sharks, while other imitators succeed when guided by a knowledgeable shark, but do not remember and fail when they are alone.

There can therefore be transmission of knowledge by imitation within flexible groups that are created and dispersed (fission-fusion). These groups, which may be made up of individuals of different ages, theoretically allow for transmission from one generation to the next. It is therefore not impossible to think that behavioural modifications could be disseminated and retained within an entire population. This takes us to the question of culture . . .

6

The Shark, Where it Belongs

2 July 2005, South Africa, off the Transkei coast. An exhausting day in the big swell of the Indian Ocean. More than eighty nautical miles on board our inflatable, more than ten hours scanning the sky in vain for . . . a school of sardines. Because, to find the fish, you have to spot the birds that feed on them. At last, there they are, high in the clouds: hundreds of Cape gannets (*Morus capensis*) turn, squawking, and suddenly let themselves drop like bombs into the ocean. We are in the middle of a crowd. Screams, collisions, foam, wings hitting the waves to tear themselves away from the ocean and back into the sky. We are hoping, if we tip over to the other side of the mirror, we will find the peace of the underwater world. But underwater, it is the same storm, slightly muffled. Millions of sardines form a compact, swirling cloud, regularly struck by birds that pierce the surface in the racket of an unlikely bombardment. Sonic vibrations hit us full in the chest from the impact of these small fry exploding. The gannets are not the only guests at the party. They have just been invited by dolphins that have been to fetch the sardines from the deep water. Like sheepdogs, the cetaceans divided and then surrounded a breakaway group of the huge shoal of

fish, before cornering it against the surface of the ocean. Each sardine, to avoid attack, tries to disappear into the heart of the crowd, and they thus create an incredible swirling ball, a monster with a thousand fins and a hundred black eyes, dressed in a chain mail of quicksilver scales. Some of them slip right into our diving suits; others come to the surface, triggering the swooping birds.

Overawed by this cauldron of life and death, we barely notice the copper sharks (*Carcharhinus brachyurus*) that come up to join in the feast. The newcomers' onslaught is a game-changer. The school breaks up. A little ball of small fry is isolated. Flashes of silver, panicked bodies, spurting scales, shearing teeth . . . The feasting raiders rip through the school again and again, until the last sardine has been swallowed. The sharks, replete, finally take off, exhausted from overeating, leaving a void behind them and the illusion that they have wiped out the whole sardine population. But twenty metres below, massed against each other, millions of other sardines form a shoal that seems to have no beginning and no end, as if what the predators took had in no way affected the stock. And in fact, this multitude persists and sardines are still as innumerable as the ocean itself!

Small fry, master of the shark

Contrary to what one might think, the number of adult sardines does not depend on shark consumption, or even on how many the totality of the predators harvest. These deductions only marginally affect the populations of small planktivorous fish.[1] Ocean currents are what naturally regulate[2] their population. Currents are the grand masters of the game of life in the sea; they are determined by winds, temperature and salinity of the water, regulating the survival of eggs and larvae, which are extremely sensitive to variations in the physical environment. Currents also transport the nutrients necessary for the development of planktonic algae, the basis of all oceanic life. Sardines are directly dependent on the quantity and variety of this plant plankton and the microscopic crustaceans that feed on it. It is therefore the abundance of food that determines the abundance of sardines, which in turn determines the survival of large predators, not the other way around. The regulation of populations always takes place from the bottom to the top of the food chain.

When sea currents do not enrich the surface waters with nutrient salts, as happens during 'El Niño' periods,[3] planktonic algae production is low and the annual reproduction of fry (sardines, anchovies and anchoveta) is poor. Populations collapse, and the large predators starve to death or are unable to feed their young. Sharks, like other predators, are put in their rightful place: they depend on their prey. It is the anonymous, insignificant animals and plants in the plankton that regulate the populations of birds, dolphins, sea lions, whales and sharks.[4]

The small appetites of sharks

What is the real impact of shark predation on their potential prey? What do they eat? I will not discuss the diets of sharks here, as they are so varied: some filter plankton, others feed on worms and shellfish, others on fish, still others on carcasses. I am interested in the quantities taken and the influence that sharks have on the ecosystem.

Let's focus on our fantasy-sharks, the super-predators that give people nightmares: the great white shark, the tiger shark, the shortfin mako shark (*Isurus oxyrinchus*), the sand tiger shark and the bull shark.

First, all honour where honour is due. 'Jaws' are considered the apex of the apex predators. Having encountered them often on dives, I can testify to one thing: the great white shark, even when hungry, will never pounce on prey or a carcass without having a good look. It is incredibly cautious and takes its time. When it approaches slowly in the deep, it is often accompanied by a host of Pacific jack mackerel (*Trachurus symmetricus*), which do not seem to be on their guard. In the waters around Guadalupe, I was even lucky enough to witness a carefree sea lion dancing under the nose of a five-metre great white! In fact, this shark eats very little. The energy needs of a five-metre white shark, equivalent to one tonne, are three kilos per day. One young sea lion is enough for three or four days.[5] This daily ration is seven times less than that of a dolphin, which is four times lighter. Like many sharks, the great white is happy to scavenge. When it finds the body of a large cetacean, it feeds on it, swallowing about fifteen kilos of its very fatty flesh with each mouthful.[6] This amount of food is enough to live between six and twenty-two days without eating.[7] Since a one-tonne shark can gobble up to a hundred kilos, it will fast for several months while it digests this gargantuan meal. In fact, most sharks must have an empty stomach before they feed again and they digest very slowly.[8] By comparison, our big trawlers catch 200 tonnes a day. The most monstrous, the *Atlantic Dawn*, has an official freezing capacity of 400 tonnes per day.[9] In just three or four days of fishing, this 140-metre vessel catches the equivalent of what the entire white shark population in Australian waters eats in a year – about 1,500 individuals.[10]

The mako shark, so fast that it can hunt swordfish, is believed to have by far the highest energy requirements of any shark, along with its cousin the salmon shark (*Lamna ditropis*). An adult shark weighing about 250 kilos would eat an average of two kilos per day,[11] the equivalent of four or five mackerel or squid. But it can just as easily gorge on shellfish or catch a swordfish and fast for days afterwards.[12] This is a far cry from the phantasmagorical man-eating mako sharks of the film *Deep Blue Sea*.[13]

To optimize the health of aquarium sharks, keepers feed them the equivalent of 2 per cent of their body weight per week.[14] Of course, this depends on the species, weight and age. Growing juveniles eat proportionally more than adults. Pregnant females also increase their intake. But on the whole, they eat a few hundred grammes per day. For example, a one and a half metre long grey reef shark (*Carcharhinus amblyrhynchos*) eats an average of 200 grammes per day. This amount is not proportional to size: an Atlantic nurse shark (*Ginglymostoma cirratum*) measuring two and a half metres and weighing eighty kilogrammes will eat 270 grammes per day.[15]

Prey on the alert?

In unspoilt places, where life seems as exuberant as at the dawn of the world, we get a glimpse of the reality of relationships among species. Around the deep-sea mountains that have been spared from fishing, the abundance is breathtaking. Here, teeming coexistence is the rule, and life seems to go on quite peacefully around sharks.

North Horn, Osprey Reef, Coral Sea, 15 November 1987. The school has no beginning and no end. I am caught in a tornado. I can no longer see my diving companions, Michel Deloire and Yvan Giacoletto. I can no longer see the vertiginous reef cliff and its shimmering corals. I can't even see the blue of the ocean any more. In front of me is a dragon with a steel carapace whose every scale is a fish. I am engulfed by a river of metal. The fish are so tightly packed that I can no longer distinguish them from each other. Fins mingle with gills, eyes are coming out of bellies. There is nothing but a shimmer of silver, shades of grey and glints of mercury struck by the sun. These are bigeye trevally (*Caranx sexfasciatus*), the largest of which measure nearly a metre; are we counting in thousands? Tens of thousands? More? How do I know? I swim straight ahead like drilling through a wall. And when, at last, I find the reef, I am again submerged by a swarming mass of green humphead parrotfish (*Bolbometopon muricatum*), as powerful as buffalos, hitting the coral with their big heads. These giants don't even change course when they pass the dens of the monstrous potato groupers (*Epinephelus tukula*). Everywhere waves of yellowtail fusiliers, streams of surgeonfish, thick curtains of unicorn fishes and snappers that enchant the seascape. Such exuberance! Life overflowing in all directions.

In all, more than sixty species were recorded in two hours of diving. And sharks were among them, but almost insignificant. Yet there were lots of them. I counted about twenty in my field of vision: coral sharks, grey sharks and two whitetip sharks that exceeded two and a half metres. A few curious ones approach me. They have to dig a tunnel through this wall of fish which will hardly move away. A moment of light, immediately masked by the living curtain that closes behind them. A few vigilant white trevallies accompany a coral shark, perhaps to make sure it moves away from the school. They close in on its flanks and attack it viciously with their fins. The prey sends the predator on its way.

We stay four days on this isolated and preserved reef, barely enough time to realize how generous and overabundant a pristine sea can be. It gives us time to wonder about the relationships between sharks and other fish – all are potential prey.

When a shark approaches, the fish know perfectly well whether or not it is on the hunt. They change their behaviour accordingly, paying more attention when it is around. They move away to keep a safe distance. They may even stop eating.[16] But once the shark's nose passes by, they resume their activity. When the shark is not on the hunt, which is most of the time, the fish go about their business without the slightest concern about it being there.

An ambiguous alliance

Sometimes, even small fry take advantage of the presence of a large shark to deter a horde of predatory tuna, at the risk of having their ally turning against them.

18 November 1985. In the open sea off Cuba. Lat. 20° 53′ north; long.: 79° 10′ east.

A warning is issued for Hurricane Kate. This will be a last launch before *Calypso* takes shelter in the port of Cienfuegos. Rough seas; strong wind. Rain is sputtering on the surface. There are white caps on the waves. We are swimming in thirty metres of water. A gigantic figure looms above us under the turbulent skin of the ocean. A twenty-metre monster, its body a constantly changing coat of glittering sequins. We approach in disbelief: millions of sardinella and anchovies are enveloping the body of a whale shark. This giant is literally engulfed by small fry pressing against it. I have never seen anything like it; a shining cloud animated by an incessant concentric movement, as each sardine seems to want to be as close as possible to the shark. I suddenly understand why: at the back of the caravan, a horde of yellowfin tuna and bonito follow at a respectful distance. The small fry have found an ally to deter them. For the moment, this strange teamwork seems to scare off the predators. I marvel at the stratagem. Suddenly, as if tired of dragging its hoopla, the nonchalant giant dives vertically and disappears into the darkness of the deep. The bonito and tuna immediately go on the attack. The small fry form one body. The shoal contracts, like a cumulus storm cloud, and rises up

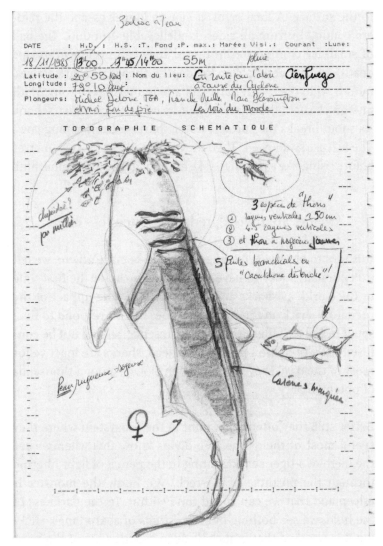

Fig. 21 A whale shark devouring the fish who had taken refuge next to it. 18 November 1985, off Cuba. Author's diving log.

to the surface. A fatal error; it can no longer escape the raiders coming in from all sides. Unbelievable eruption. The ball of small fry contracts even more. They can no longer escape death. And, emerging from the depths, the whale shark arrives, mouth wide open. The tuna call a halt and make room. With one great suck, the shark engulfs part of the sardine school. Its snout breaks the surface, water spurting from gaping jaws. The fish are swallowed. Then the former ally sinks a few metres before rising like a piston to take another gargantuan mouthful.

Sharks are there; 'predators' much less often

It is rare to observe a feeding shark. On board *Calypso*, we felt exceptionally lucky to have seen such a meal for the first time in the world, as sharks are usually resting. The impact of the predatory shark on the ecosystem does not correspond to time spent in a place. Therefore, 'shark presence' should not be confused with 'predation pressure'. Indeed, sharks are inactive for most of the day. They sleep or swim slowly, among thousands of potential prey.

Better still, they often don't hunt in the ecosystem where they spend most of their time. We divers know this when we see the teeth of a tiger shark glinting in the pencil of light filtering through the porthole of a wreck. We know the monster is asleep and that we can spend time with it. In the darkness of the hold, we see nothing but the dozens of sharp fangs, sticking this way and that in its half-open mouth. One of the most frightening mouths one can ever see. The monster reveals itself gradually. Three metres long, with a long, pointed snout that extends a massive, antediluvian, almost ungainly silhouette; short pectorals, a very heterocercal caudal.[17] Not particularly elegant. But its black pupil is surrounded by gold, with a hypnotic effect. Capable of lightning accelerations when hunting, the tiger is nonchalant. Holding my breath, I approach it calmly and respectfully so as not to break this tranquil moment. I am so close that I can count the pores of the Lorenzini cells piercing its nose. It remains indifferent during the forty-five minutes of our dive. It will remain so long after we leave the wreck. And the dozens of yellow snappers that flit around its snout are well aware of this. But when night comes, it will wake up. Slowly, it will leave its den far behind and plunge deeply, up to 150 or 200 metres, to go on the hunt.

Home-delivered meals

On the other hand, sharks can influence reefs that they do not frequent. This is what the scientists who accompanied the explorer Laurent Ballesta on his incredible dives at Fakarava, in French Polynesia, have brilliantly demonstrated. The atoll passage is frequented by 700 grey and coral sharks. This abundance of predators exceeds the low productive capacity of this small area. It looks like it is an inverted trophic pyramid.[18] This is obviously impossible, and incomprehensible, given the scale of the passage. But this passage welcomes, in addition to its usual residents, a large number of visitors. In particular, the fish that come to reproduce there. Surgeonfish, parrotfish and no less than 17,000 groupers (*Epinephelus polyphekadion*) gather in the large passage for the first full moon of the austral winter.[19] This is a considerable biomass of reproducing fish, coming from all over the world, concentrated in a very small area.

So, these sharks, which seem overabundant for the area, are actually harvested over a much wider area, since the groupers come from very distant places. In fact, the groupers are 'importing' everything they themselves have eaten on other reefs.[20] As a result, although they do not move, Fakarava sharks indirectly exert predation pressure on reefs they do not visit.

And yet this pressure remains very moderate: the 700 sharks consume about 150 kilos of fish per day, or about fifty groupers weighing three kilos.[21] At the end of the three-week reproduction period, barely a thousand of them (i.e., 6 per cent of the 17,000) would have perished under the sharks' teeth, if the latter had only eaten groupers. This is far from being the case, as they consume many other species: surgeonfish, moray eels, trevally, unicorn fish, parrotfish, etc.

Prey as a baby, large predator as an adult

Measuring the influence of sharks on the ecosystem is all the more difficult because, over the course of its life, the shark changes not only its diet but also its ecological niche.[22] Before becoming a 'big super-predator shark', the young shark of any species is easy prey, especially for its elders, who are the main threat.[23] Even the newborn white shark, which measures 1.2 metres, is at the mercy of its brothers, cousins and cetaceans such as killer whales. It has neither the protection of its parents nor that of a cohesive social group, as do dolphins or sperm whales.

Very opportunistic, the young great white feeds mainly on fish until sexual maturity (3.4 metres), which marks a radical change in diet. It will then prefer marine mammals.[24]

The young tiger shark is a mesopredator that consumes fish, crustaceans and especially molluscs.[25] It is only as an adult, when it exceeds three metres, that it attempts to attack large herbivores such as green turtles and dugongs (*Dugong dugon*). However, its attempts are rarely successful. The prey does not take kindly to this, as shown in an astonishing video shot by a flatback turtle (*Natator depressus*) on which biologists had set up a camera. Attacked by a tiger shark, the turtle is very aggressive and does more than defend itself. It tried to bite its attacker and then escaped.[26]

The influence of sharks on these herbivores is therefore less in the number of animals consumed than in the state of alertness brought about when it prowls over the seagrass beds.[27] Thus the repeated claim in popular science literature that sharks protect seagrass beds and make the oceans more resilient to climate change by limiting the number of grazers such as turtles and dugongs is a bad oversimplification.[28] The impact of sharks is paltry, except for those populations that are in danger of extinction . . . because of human activities.

Cascade effect, from theory to reality

In order to draw attention to the unacceptable mass destruction of sharks, some people are trying to outdo each other. They claim that sharks support oxygen production.[29] To this end, they invoke the cascade effect,[30] well known to Anglo-Saxons as the 'top-down effect', which has been very well documented in terrestrial environments, in particular the positive impact of the return of wolves on the diversity of plant species in the forest.[31] Are these very simple and remarkable examples, such as that of 'large carnivore, large herbivore, young tree consumer', applicable to the marine environment, where food webs are infinitely more complex? According to this principle, apex-predator sharks would be the regulators of mesopredator populations (groupers, trevally, barracudas, moray eels). Without sharks, these mesopredators would swarm. They would become so numerous that they would wipe out the small planktivorous fish (fusiliers, sardines, etc.). The disappearance of the latter would lead to a proliferation of their prey: small crustaceans that graze on plant plankton! In short, by domino effect, if sharks disappear, small crustaceans will be so numerous that they will graze on all the microalgae in the plant plankton that provides two-thirds of atmospheric oxygen.[32]

Sharks, the great regulators of life on Earth! How did we get here?

A 'romanticized' idea of the marine food chain

This reasoning is based on a doubly simplistic representation: on the one hand, a determinist vision of living beings which holds that each creature has a function, and on the other, the extreme schematization of the notion of the 'food pyramid'.[33]

This simplification is largely due to the study by Professor Ransom A. Myers, who demonstrated that the disappearance of sharks had led to a proliferation of their prey, cownose stingrays.[34] Too many of these rays would wipe out the scallop population on which they feed, impoverishing scallop fishers. Setting up rays as the perfect scapegoat gives the fishers a green light to overexploit scallops, and the shark is set up by scientists as a great environmental manager. Everyone had something to gain: environmentalists, scientists, fishermen. In short, the story was such a good one that no one wanted to discuss it any more. At last, we had the 'scientific proof' of a cascade effect operating as smoothly as the cogs in a watch.

But it doesn't work that way, not for scallops, nor for seagrass beds, nor for oxygen. It is the algae that are the masters of life, not those who eat them.[35] On the other hand, the influence of large predators on animals, such as fish, whose reproductive strategy is of the 'r' type, which relies on the release of millions of eggs into the environment, is negligible compared to the influence of environmental factors.[36] In the sea, there is no such thing as a linear 'food chain', as it is represented with successive cogs. On the contrary, it is a network of formidably complex relationships, with hundreds of interactions among all the organisms, defying our hierarchy of trophic levels. The smallest sardine can feast on its predators, mackerel and tuna, when they are still in their larval stage.

Sharks are not picky eaters

A shark is incredibly opportunistic. It feeds on a wide variety of species from all trophic levels.[37] So much so that two sharks from the same population may have very different diets.[38] This variability, which also leads them to feed in different ecosystems, taking advantage of the slightest opportunity, considerably reduces their real impact on a specific species and, as a result, the much-invoked cascade effect, especially as large predators only take from the adult fringe of the population.[39]

Diet analysis of seven pelagic shark species shows that prey diversity is greater within species than between species.[40] Sharks graze from every available outlet.

Even better, the study of the stomach contents of five shark species (copper, dusky, mako, hammerhead and thresher) shows the incredible diversity of their prey, which belong to nearly forty taxonomic groups and occupy up to ten trophic levels.[41]

Finally, the cascade effect is all the more difficult to measure as marine species are very redundant. Shark prey (groupers, jacks, barracudas) can eat very young sharks and feed on the same prey as their predators.[42] I have often seen black jacks (*Caranx lugubris*) accompany coral sharks on their nightly hunt. Not only were the jacks never bothered, but they ate the prey the sharks had found, right under the sharks' noses.

A brief aside on the importance of redundancy

This highly flexible redundancy among competing species con-tributes to life's amazing resilience.[43] It is all the more important to remember this because some economists, who have become aware of the economic services provided by nature, want to 'optimize' these services by ranking species: those that provide the best service would be preserved, while the others, which are more or less redundant, could be eliminated, forgetting that it is precisely diversity that underpins the general balance of life.

Sharks live as best they can in a very complex ecosystem. They act more as a mesopredator than as a super-predator whose role is to regulate mesopredators, as popular imagery would have it.[44]

This is also what Laurent Ballesta's team found out as they worked on the interactions between sharks and groupers in the Fakarava channel. The 17,000 groupers, which are in the midst of reproducing and therefore more inattentive than usual, should be an easy source of food for the sharks. But this is not the case: 'Although the sharks clearly benefit from this gather-ing, they are very opportunistic and prey on more than forty different species. They feed on the entire food chain. In the end, the sharks take very few groupers compared to the total number, so they do not affect reproduction.'[45] Groupers are secretive and hard to catch. On the other hand, some plank-tivorous or grazing fish, such as surgeonfish and unicorn fish (*Acanthuridae*), panic when a shark passes by, triggering an attack. Thus sharks, which are supposed to regulate groupers, actually devour many more small grazers and planktivores, i.e. the very ones they are supposed to be protecting from groupers!

A place, not a role

The other reason we exaggerate the importance of sharks in the ecosystem is semantic: the confusion of the words 'role' and 'place'. When authors choose the word 'role', it induces a bias: it focuses the research on the 'top-down' impact of the predator on other species, neglecting the influence of the latter on sharks. Worse, the scientist tries to highlight more than just a function: a 'mission' in the service of a 'project'. The result is that in the field, the observer is determined to show how the shark, a super-predator, performs its 'job' of regulating the balance and 'guaranteeing' the health of the ecosystem. He or she attributes it with the objective of eliminating the sick, the weak, the harmful and the supernumerary, blinding themselves to the reciprocal relationships that actually bind species together. This deterministic view of nature is in response to a dominant Judaeo-Christian culture, which is all the less likely to be questioned because it justifies the hierarchy of our society built on the primacy of the strongest. And it responds to a prevailing social Taylorism: one job per person. This leads to the now paradigmatic idea that each animal has a predetermined function, if not a divine mission.

Even documentary filmmakers focus on the strongest predators. It energizes the narrative and keeps the viewer's attention. To justify this editorial choice, the directors explain in the commentary that 'as a super-predator, the role of the shark is to "regulate" the ecosystem'. The truth-effect of images contributes to this deterministic and caricatural vision of nature, without anyone questioning the subjectivity of the choice of images or their representativeness. The cameramen are careful not to say that it took them weeks to shoot these predatory scenes or that, if they had to show the public sharks in their everyday lives, they would have shown them sleeping or swimming peacefully amidst myriads of fish.

Living things are not boxed in by the simplistic missions we attribute to them. Life does not have an objective. Nature is not deterministic. It is. It grows, multiplies, diversifies and becomes more complex through species associating. But no species has a 'role' to play in the great cycle of life. Each species takes the 'place' it can, with the assets it has, in relation to all the others, even if it is not in direct interaction with these others.[46]

Opportunistic cooperation

Sharks need others, often their own kind, with whom they nevertheless remain in competition. This is what Johann Mourier and Pierre Labourgade have shown by studying the affiliated hunting of coral and sandbar sharks. Although each shark hunts for itself, and there is no consultation, their different tactics complement each other, improving the success of each. In this exercise, grey sharks take advantage of the coral sharks, which are more agile in reef crevices. Grey sharks are open-water hunters and do not like to venture under coral reefs where sleeping fish are huddled. These hidden prey represent barely 18 per cent of their menu when they hunt alone. On the other hand, when they follow their cousins, who are better at finding fish in crevices, they represent 80 per cent of their meal.[47]

I observed the same opportunistic cooperation between predators during the Sardine Run[48] in South Africa. Copper sharks, jostled by dolphins (*Delphinus capensis*), with Cape gannets flying ahead, are only the companions at a Dantean banquet with cetaceans serving up the sardines they have herded. Rather slow predators, they clearly take advantage of the attacks launched by the dolphins on the outsides of the shoal and the diving gannets that panic the sardines. The sharks depend, in a way, not only on their prey, but also on other predators that help them succeed. The more predators there are, and the more varied the species, the more each individual eats.[49]

A lone shark has little chance of catching a sardine. It creates a vacuum around itself, with the shoal opening in front of its snout and closing again after it has passed. As Paul Valéry might have said: 'A shark alone is always in bad company'.[50] On the other hand, simultaneous attacks, from dolphins on the sides and from boobies from above, prevent the sardines from moving away from the shark, so it can bite into the panicked mass. Far from being the king of the oceans acting at his own pleasure, on a whim, the shark, like all living things, is part of an incredibly complex web of relationships. It depends on each of them.

In charge of the health of the ecosystem?

The shark is also supposed to have a 'sanitary' role, eliminating the weak and old in order to guarantee the good health of the population. Observation shows that the shark eats what is within its mouth's reach, most often fish in good condition, caught simply at the wrong time and place.

During his twenty-one weeks in the Fakarava Channel, the explorer Laurent Ballesta also filmed a large number of crippled groupers that had survived shark attacks. These wounded groupers, with fins or even parts of their bodies amputated, seemed to heal very quickly, and even come back the following years to reproduce.[51] Paradoxically, not only did the shark not eliminate them, but it caused their infirmity as well.

On *Calypso*, in November 1987, we spent five days diving during the green turtle spawning aggregation at Raine Island in the Coral Sea. Nearly 60,000 of them come here to mate and lay eggs. It's a gruelling ordeal for these marine reptiles as they drag themselves across the sand to lay their eggs as high up on the beach as possible. This annual gathering is a delight for the great tiger sharks that we came to film. I was astonished to see that many turtles had had a leg or even part of their body bitten off. They mated like the others, without worrying about their handicap. Does this mean that tiger sharks, opportunistic scavengers, are not great hunters, and that, even when wounded, their prey manages to escape them? The same conclusion as for groupers: on Raine Island, it is sharks that cripple the most.

Moreover, if it were necessary for a predator to eliminate the weak in order to maintain the health of a population, how would this be done by super-predators that have no predators, such as killer whales? Indeed, how is it done by all those species (whales, elephants, lions) that have very few predators except in the early years of their lives? They are at the mercy of the real super-regulators: weather, bacteria and viruses.[52]

Almighty microbes

Sharks are renowned for their resistance to viruses and bacterial infections and, above all, for their astonishing ability to heal.[53] This resistance is thought to be due to high genomic stability through gene redundancy and, more specifically, to positive selection during phylogenesis of genes involved in scar tissue healing.[54]

It is even a popular belief that sharks do not develop cancer. This superstition is carefully nurtured in the United States by the pharmaceutical industry, which is eager to market anti-cancer food supplements based on shark cartilage. Pseudo-scientific bestsellers as popular as the film *Jaws* even extol the virtues of such therapies.[55] Yet, like all living creatures, sharks develop monstrous cancers. And sometimes, divers see sharks horribly deformed by huge tumours. These cancers even reach the famous cartilage that is supposed to prevent the development of disease in humans.[56] Sharks are not only susceptible to cancer, they are also vulnerable to all sorts of other diseases. For example, the bacterial infection *Streptococcus agalactiae*, which causes horrible epizootics in fish, affects sharks with the same virulence.[57]

A pantry for parasites

Like all living creatures, each shark is part of an ecosystem in which, far from being the king, it has to put up with a lot. Many small animals do not see it as a predator but as a pantry. The shark is in fact the habitat of a formidable community of parasites that take advantage of everything: mucus, skin, blood. Up to 3,000 parasites have been counted on the body of a single blue shark, on its gills and even in its nostrils.[58]

The great white sharks I encountered were covered in small parasitic copepod crustaceans which make life a misery for them.[59] Some, such as *Nemesis lamna* or *Pandarus satyrus*, can infest the gills to the point of killing the animal.[60] *Ommatokoita elongata*, another copepod, gets into the cornea and feeds on the eye. Most Greenland sharks are infested and blind.

Copepods are not the only ones to torment sharks: *Anelasma squalicola*, a cirripedean crustacean, sucks the flesh from the reproductive organs of its preferred host, the ninja lanternshark (*Etmopterus benchleyi*). The worms are not to be outdone: *Trilocularia eberti*, a cestode, infests the intestines of sharks of the genus *Squalus* which, on the other hand, are the favourite prey of the great white shark.[61] Being a shark is definitely not easy . . .

Fig. 22 Parasitic copepods (*Pandarus satyrus*) sucking on the tip of the tail fin of a great white shark.

Fig. 23 The parasitic remora fish (*Echeneidae* sp.) fixed on the back of a great white shark.

Although the skin of sharks is extremely tough and covered with denticles, it cannot resist the dozens of small teeth in the mouth of the sea lamprey (*Petromyzon marinus*). This large, primitive, anguilliform vertebrate hooks into its victim's muscle, which it digests with the help of a highly effective anticoagulant.[62]

Each parasite has its own tactics. The ferocious dwarf lantern-shark, a small shark of fifty centimetres, surprises those who venture into the deep, including the great white shark. In a lightning attack, it bites down, spins around and tears off a huge mouthful of flesh and disappears into the darkness of the abyss as quickly as it appeared. It leaves its prey with a bloody crater about ten centimetres in diameter.[63]

It is these parasites, microbes and fungi that are the true predators of the ocean.

Sharks' allies

Fortunately, the ocean is not just full of nasty profiteers who mistreat the poor shark. Its closest allies are also microbes that protect it from other pathogenic microbes and help it to heal without infections.[64] The micro-organisms that are most essential to the shark are found in its digestive tract. Dozens of families of symbiotic digestive bacteria have been identified. They go by the gentle names of *Cetobacterium* sp., *Proteobacterium* sp., *Vibrio* sp., *Proteobacterium ribotypes*, *Actinobacteria*, *Firmicutes* (*Clostridium* sp.), *Fusobacteria* (*Cetobacterium* sp.), *Proteobacteria* (*Campylobacter* sp.). Some produce enzymes so powerful that they digest the indestructible chitin of crustacean shells.[65] Co-evolution has made the 'shark-micro-organism' community so inseparable that one wonders who is more indispensable: the bacteria or their host? In other words, is the shark merely the shell-habitat of a formidable community of bacteria?[66]

Carers to the rescue

The ocean is also home to legions of little carers who come to the rescue of sharks in need. These seemingly insignificant cleaners were unknown until the advent of scuba diving, as their behaviour can only be observed in their underwater home. But you have to sit back and pay attention to discover these fish and shrimp waiting for their patients at the bend in a rock. Then you have to develop a great deal of discretion so as not to frighten those giants that come to check in at the infirmary.

9 March 2001, 4 p.m. El Arrecife 2, north-east coast of Malpelo Island (Colombia), twenty metres deep. The water is so rich in plankton that visibility does not exceed ten metres. I stand motionless on a rocky ledge dotted with pocilloporid coral heads. One of them is besieged by twenty or so leather bass (*Dermatolepis dermatolepis*) who launch unsuccessful assaults against the coral fortress. It is not the madrepores that these formidable plunderers are after, but the small fry that have taken refuge there. The bass are organized in a pack, and patient. While one attacks, another waits on the other side, ready to swallow an unwary fish trying to escape. The charges follow one after another and are violent to the point that the bass hurt themselves against the coral branches. No luck, the besieged are not dislodged. All this action attracts a few trevallies who would like to take advantage of the final assault . . . But there will be no final assault. The bass give up and leave in search of easier prey. The tension subsides. The trevallies stay put, nose to the current. And, surprise, the little fish, hiding in the midst of the coral branches, comes out all debonair! It is a young hogfish (*Bodianus diplotaenia*) which, without a moment's hesitation, rushes towards the first trevally which is floating motionless, slightly tilted upwards. With its mouth wide open, it seems to put itself on offer. The little hogfish enters its mouth and begins a meticulous pest control. It pecks, it pulls, it eats infinite things that I cannot see, but which seem particularly to bother the languid beauty. Two other trevallies have straightened up and are waiting for the services of the cleaner fish, which already has a lot to do with its first client. Three yellow butterfly fish (*Johnrandallia nigrirostris*) arrive to groom the gills of the other patients. All is calm and peaceful. How far away the turmoil of the previous minutes seems!

However, in the distance threatening silhouettes are emerging. A dozen hammerhead sharks are approaching, led by a beautiful female of more than two and a half metres. She is badly wounded above the right pectoral fin. A deep whitish tear has cut three gill slits. Bites inflicted by a male during mating? The troop passes five metres above the cleaning station without anyone getting upset! The sharks disappear, only to return two minutes later, this time swimming close to the bottom. I hold my breath and try to blend into the rock so as not to scare the very shy hammerheads. They will pass within a few metres. The large female swims very slowly. Her flat head swings from right to left, regularly uncovering the ugly wounds. The hogfish and butterflies immediately abandon the meek trevallies to inspect the shark's wounds. They have little time, as the shark, which cannot stop swimming, moves away from their coral refuge.[67] They hesitate, but do not risk exposing themselves. They return to work on the waiting trevallies' parasites. But the female hammerhead comes back so that they can complete their first inspection! She treats herself to two more rounds of care.

Butterfly fish cleaning a hammerhead shark
(Galatée Films).

Like the hammerheads, the very discreet thresher shark rises daily from the ocean depths to meet these carers. To announce that it has come as a patient, not a predator, this deep-sea runner displays a panoply of stereotypical contortions, swimming in large concentric circles so as not to stray far from the cleaning station. The bluestreak cleaner wrasse (*Labroides dimidiatus*) can then gobble up the parasitic worms that infest their imposing patient causing all sorts of diseases.[68]

Grey reef sharks, on the other hand, hold themselves almost motionless, tilting their heads slightly skywards, mouths gaping. They allow the wrasse to work between their teeth, to rummage through their gills, to enter and exit at will through the gill slits. Never, ever, will a cleaner at work be eaten. Every shark, whether offshore or resident, benthic or pelagic, knows the vital importance of these little allies, it's like Jean de La Fontaine's fable 'The Lion and the Rat'.

Full- or part-time cleaner

There are over 130 species of cleaners, both crustaceans and fish.[69] Some are strictly specialized – such as the wrasse, which as an adult is barely ten centimetres long. Others are only active in the juvenile stages, such as the hogfish (*Bodianus* sp.). Others, such as butterflyfish or angelfish, only work part-time. Finally, some fish are opportunistic cleaners, like the pacific jack mackerel (*Trachurus symmetricus*), which I observed cleaning the wounds of a huge female white shark in Guadalupe.

Cleaners are the indispensable 'doctors' for the inhabitants (fish, sharks, turtles, mammals) of all marine ecosystems. All of them, without exception, owe their good health to these insignificant creatures. The importance of their care for the balance of ecosystems is probably far greater than that of super-predators. Peter Waldie's team, which conducted an experiment on eighteen coral reefs for more than eight years, has shown that without their cleaners, the ecosystem is disrupted: its species diversity drops by 23 per cent. The number of resident fish of all species collapses by 37 per cent. Weakened or killed by parasites, they are noticeably smaller. Juveniles do not settle on a reef without carers (67 per cent less). Even adult visitors are less numerous (23 per cent) and their overall species diversity has fallen by 33 per cent. This decrease is particularly noticeable for herbivores (66 per cent), which give free rein to the algae that overwhelm the corals. It is true that by inspecting an average of 2,297 patients and gobbling up some 1,218 parasites every day, the little wrasse is a very unusual carer.[70]

However, as with doctors, there are more and less popular cleaners. The cleaning stations are unevenly frequented. It is not the location that is to blame, but rather the sensitivity of the cleaners. The cleaners prefer the mucus of the fish to the parasites and sometimes carelessly gobble up the mucus of their patients. Clients also shun facilities with awkward cleaners who rip off skin along with the parasites. Cleaners therefore have to be gentle if they want to eat copepods, dead skin and mucus with gusto.[71]

In some places, the cohorts of cleaners are so numerous that patients are literally assaulted! In October, in the Cocos Islands marine reserve (Costa Rica), dozens of butterfly fish literally assault hammerheads to clean their wounds such that, at times, the sharks are so annoyed and powerless that they frantically contort themselves to get rid of them.

Prey in rebellion

In these places of overabundance, the sharks are ordinary actors, lost in the midst of the crowd, often pursued by their prey. A troubled interaction. A deliciously ambiguous relationship, symbolized by the schools of rainbow runners (*Elagatis bipinnulata*) that harass coral sharks in order to scratch on their rough skin. The sharks try in vain to flee the assaults of these intruders.

Coral shark attacked by rainbow runners
(Galatea Films) and a stingray attacked by a group of
bigeye trevallies.

One of the funniest sequences in Jacques Perrin and Jacques Cluzaud's film *Oceans* is at the expense of a blue shark, which is annoyed by a remora fish that suddenly takes refuge in its gill cavity, as if it were home.[72]

On 3 August 2000, during the 'Great White Shark' expedition, we were scouting off Dyer Island, south of the Cape of Good Hope. This island is home to hundreds of fur seals (*Arctocephalus pusillus*) and a few Cape penguins (*Spheniscus demersus*). The latter are famous hunters of sardines, which are very abundant in these waters. One of them, knee-high to a grasshopper, dives into the waves just as a four-metre white shark approaches. The shark immediately spots the bird swimming on the surface. It makes a wide turn, disappears and reappears in the penguin's wake. Its rapid swimming is unmistakable. But the most violent attack is launched . . . by the penguin. The bird strikes the unfortunate shark's head with its beak and the shark flees while the going is good.

Three years later, on 30 April 2003, I dived again at the same spot. The sharks made me wait, then one of them finally appeared in the green water. It is a male. I hold my breath and hold still so as not to frighten him. I adjust my photographic frame to get the best possible angle and get the kelp in the scenery.[73] The shark, somewhat backlit, is about to pass between the large fronds of seaweed when, without warning, three torpedo-like forms are in pursuit. I barely have time to see the fur seals angrily biting the poor shark as it tries to escape them in the murky water.

'Jaws', a prey like any other

Even 'Jaws' have to contend with others, especially with other large predators.

In 2000, during my first expedition as scientific director of the Deep Ocean Odyssey programme, it was estimated that 500 great whites were living between the Cape of Good Hope and the Agulhas Cape, 150 kilometres apart as the crow flies. More than 70,000 fur seals, 17,000 Cape penguins and countless small sharks, dogfish (*Poroderma* sp.), provided all these predators with abundant food. This area was their paradise and for us divers this was the best place to encounter them.

Fifteen years later, this paradise has turned into hell to the extent that in 2017 and 2018, no white sharks were seen. At the same time, scientists have noted a gradual increase in the number of bluntnose sharks, another fearsome predator that no one had seen before.[74] The bluntnose sharks are more daring and aggressive, hunting very efficiently in groups; could they have caused the disappearance of the white sharks? Could it be that the white sharks could not stand the competition? Or did the bluntnose sharks simply take advantage of the disappearance of the white sharks to occupy the ecological niche left vacant? No one had time to answer because a third thief, or rather two thieves, two killer whales named Port and Starboard, appeared in the area. Perhaps they were justices of the peace. In 2017, seven disembowelled great whites were found on the beaches with their livers eaten. These signs seem to point directly at the cetaceans, which are known to eat only the tastiest organs of the sharks they prey upon.

Could these two killer whales have dealt with several hundred white sharks in just two years? It is unlikely. Although their presence certainly disrupted the presence of great whites, the orcas should have caused the disappearance of the bluntnose sharks they also hunt. And this has not been the case.[75]

Are not killer whales, like bluntnose sharks, good scapegoats to divert attention from the real cause of the disappearance

of great whites: industrial bottom longline fishing? Not the
fishing of white sharks themselves, but of their prey: sharks
of the genera *Galeus* and *Poroderma*, which account for 60
per cent of their diet. Astronomical quantities of these small
sharks are exported and consumed in Australia as fish and
chips. The collapse of the South African populations is such
that scientists have launched an awareness campaign to stop
the consumption of shark with chips.[76]

In 1988, during an expedition to Papua New Guinea that for
the first time brought together the two boats in the Cousteau
team, *Calypso* and *Alcyone*,[77] my diving buddies Clay Wilcox
and Louis Prézelin were greeted by two orcas. The two cetaceans had suddenly dived down to chase a two-metre grey
shark, then a manta ray, which they had taken the time to
devour in front of Louis' camera. Thanks to increased dive
observations, we now have a better idea of the pressure that
large cetaceans exert on sharks.[78] It is also easier to understand
why white sharks around the Farallon Islands in California
prefer to cautiously move away from elephant seal colonies
(*Mirounga angustirostris*), their favourite hunting ground,
when orcas approach them.[79]

Sharks as almost indispensable to the ecosystem . . .

So, super-predator or not, the shark is not the master of the
game, it is only one element in a network of close interdependencies. However, there is a moment in the shark's life when
its contribution to the ecosystem is essential. It is when, as a
corpse that has dropped into the deep sea, it offers itself to
the scavengers of the deep who would go hungry without it.
Without light to allow photosynthesis, deep-sea ecosystems
are deprived of plant production. They rely to some extent on
chemosynthesis by hydrothermal bacteria, but depend mainly
on manna that falls from the surface. This manna arrives in the

form of organic snow, made up of tiny plankton debris that continuously flows down from the ocean's sky. Sometimes, in the midst of this food shower, the corpse of a whale, tuna, swordfish or shark falls. Never has death been so recognized as a source of life as in the abysses, where corpses give rise to oases that last for years. The bodies of cetaceans, which are rich in fat, can even be the source of the development of chemosynthetic bacteria that support a thriving community of worms, crustaceans and abyssal fish for decades.[80] Sharks and rays, however, do not offer such opportunities, as their bodies are not sufficiently rich in fat and their cartilaginous skeletons are more quickly eaten. Nevertheless, the rare observations of shark and manta ray corpses lying on the abyssal plain show that they feed a whole fauna of large necrophagous animals for months on end: abyssal sharks, lycodes fish (*Pachycara crassiceps*) and amphipod crustacea.[81]

7

The Ocean is Their Garden

Letter to my human friends

'I was born in 1748. That year the Seine froze over in Paris. During the whole month of January it was so cold that even the olive trees died in Provence. The average temperature was −13°C. A very pleasant temperature. Unfortunately, June was a scorcher with an average of 36.9°C.[1] That same year, a little further inland, Lahore was sacked by the King of Afghanistan, Ahmad Shah Abdali, who defeated the troops of the Mughal governor, Shahnawaz Khan, on the River Ravi. The English fleet failed to take Pondicherry. And your King Louis XV signed the Treaty of Aix-la-Chapelle, ending the War of the Austrian Succession . . . Ah, yes, another thing: Monsieur de Montesquieu published *The Spirit of Law* and a Neapolitan peasant rediscovered the ruins of Pompei . . . buried under the ashes of Vesuvius for more than 1,600 years:[2] the time of my great-great-grandfather . . .

Ah, the restlessness of men!

You and I, dear humans, do not live at the same pace.

I am Eqalussuaq, the Greenland shark, but your researchers have named me *Somniosus microcephalus*. I am 274 years old, but some of my big brothers, who are over six metres long, are almost 500 years old! They would have been born at the time of your Christopher Columbus, who imagined being the first to cross the ocean . . . Our species has the privilege of having the longest lifespan of all vertebrates![3] On the other hand, I reached my sexual maturity only 120 years ago. Needless to say, at this rate, I need time and peace of mind to ensure I have descendants. Fortunately, I live away from the world, in the cold waters of the Arctic. I also enjoy the deep, cool waters (4°C) that allow me to travel to the trenches of the Gulf of Mexico.[4] Nor do I disdain the waters of the St. Lawrence, which I frequent assiduously, even when they warm up to 16°C. I am cold-blooded. I move slowly. I swim at my own pace, 2.7 kilometres per hour.

My distant cousin, the salmon shark, does even better. It too frequents the icy waters of the Bering Strait, where it remains, even in winter, on the edge of the ice pack.[5] Although the water temperature does not exceed 1°C, it can swim at more than fifty kilometres per hour and shoots its two metres and one hundred kilos out of the water when hunting. These incredible performances are due to its body temperature, which it can maintain at 20°C above the temperature of the surrounding water! It is endothermic like mammals.[6,7] This asset opens up worlds that are foreign to me, even in the warm waters (20°C) of Hawaii, thousands of kilometres from our cold world. One of them even swam 18,000 kilometres in just under two years! You see, dear humans, our space-time, as sharks, is definitely not yours . . . So, with my privilege of age and experience, please take a step back to better understand our lives. Yours truly. Eqalussuaq.'

Embracing the immensity of the shark world

How can we imagine the extraordinary odysseys that sharks undertake, far from sight, in the ocean abysses? Of course, the ancients, fishermen and naturalists, suspected such journeys, as they had noted their seasonal arrivals and departures. Marcel de Serres, the first to have studied the migration of fish, wrote as early as 1845: 'At least we see, in the southern regions of France, the passage of sardines constantly coinciding with that of mackerel, like their migrations with that of tuna and sharks.[8] This remarkable coincidence is renewed with such great regularity that a somewhat irresistible instinct must regulate the periodic journeys of these animals'.[9]

But how can one follow these peregrinations that the surface of the ocean hides from the eyes of men?

In 1990, Jean-Michel Cousteau was determined to shed light on the movements of great white sharks in the waters off southern Australia.[10] With exceptional resources for the time, his team boarded the *Alcyone* and headed for Dangerous Reef, home to the world's largest colony of sea lions (*Neophoca cinerea*), one of the great white's favourite prey. On board, Barry Bruce, a young scientist, used waterproof acoustic transmitters, which he wanted to attach to the sharks' backs, to track them from a distance. Not so easy, as he had to keep his receiver right above the transmitting shark so as not to lose it. Exhausting nights on the open sea in a tiny boat listening to the little 'beep-beep-beep'. At the time, there was no laptop computer; the journeys were not even recorded simultaneously. It was on board the *Alcyone* that the paths of the small boat tracked by radar were plotted by hand on a real map . . . Incredible handiwork, which for the first time allowed the world to follow Antoinette, a pretty female white shark, for twenty-two hours over a distance of thirty-six miles, or sixty-five kilometres. After this great beginning, and two years of study, Jean-Michel concluded: 'the sharks swim either just below the surface or along the bottom where depths average sixty-five feet . . . White sharks tend to congregate according to gender. Females were more abundant at Dangerous Reef and other inshore islands, while males were found near the Neptunes and other offshore islands.'[11]

Miniature beacons and long-distance migrations

Thirty years later, Barry Bruce and his team from the Commonwealth Scientific and Industrial Research Organisation (CSIRO) are studying the same population of great white sharks, in the same place, but with state-of-the-art miniature satellite tags. The findings are astonishing: great whites do not travel a few dozen kilometres, but several thousand kilometres. In less than five months, one female even travelled 12,240 kilometres and dived to a depth of more than a thousand metres.[12] These results support those obtained by researchers in South Africa who followed Nicole, another female white shark, for six months, over 20,000 kilometres across the Indian Ocean.[13]

Why such long journeys, when food is plentiful along the Australian and South African coasts? What about other populations? What about the population of Guadalupe, in the North Pacific Ocean, which we are exploring for the film *Oceans*?

Guadalupe, the rendezvous of lovers?

10 November 2006. No current this morning. The water is full of plankton. This is our sixth dive with the great whites. I float in the deep blue water, fifteen metres below the surface. A small, very nervous male, whose dorsal fin is severed, comes and goes without stopping. A shy female also remains a distant

blur. Nothing here to please Didier Noirot, the cameraman, and Pascal Kobeh, the photographer, whom I can barely make out when they are less than twenty metres away. Suddenly, an 'impression', which I feel rising from the bottom, under my fins. A grey line that circles slowly, then straightens and accelerates. The silhouette becomes a shape. It takes on fins, muscles, a monstrous neckline. Its pectoral fins are lowered, indicating its high nervousness. It arrives very quickly and only averts its charge a few metres away. It passes in front of my mask, chewing. I immediately notice the small clear mark at the base of the caudal fin; the white coat of her belly which rises very high, clearly isolating the darker pelvic fins, and above all the deep tears which cut into her left flank from the pectoral to the cheek. This is Lady Kathy, a large female of five metres in length. For sure, the most scarred of the many ladies who frequent this place in autumn. Scars probably inflicted by a very ardent lover . . .

Do sharks travel to Guadalupe to breed? Do the sheer cliffs of the volcanic island, which have their roots 3,000 metres deep in the Pacific Ocean, hide their secret love affairs?

White Shark Café

The answer provided by the satellite beacons is clear and, with it, there are revelations about the formidable odysseys of these ladies: yes, Guadalupe Island is indeed the mating ground for a large proportion of the white sharks of the north-east Pacific. Not only do these sharks undertake fabulous journeys into the heart of the ocean, but contrary to popular belief, the females spend much more time (65 per cent) in the open ocean than near the coast.[14]

The journeys begin in early February. Males and females disperse. They head out to sea. Some go as far as Hawaii, 4,500 kilometres away. They leave the surface zone and sometimes go more than a thousand metres deep. On this immense journey, no shark will miss a stop at the White Shark Café. A mysterious place, in the middle of the ocean, in the middle of nowhere, where sharks from the two populations of the northeast Pacific, those of Guadalupe and those of California, meet without anyone knowing why. The hydrological conditions in the minus 600 metre zone seem ideal for congregations of squid, tuna and swordfish.[15] It's only a short step from there to thinking that the White Shark Café is an exceptional roadside eatery, located on the border of the twilight zone.[16,17] In any case, after having had their fill, everyone goes their own way.

The males are very punctual and return to Guadalupe every year in August. The females, on the other hand, skip a turn. They travel for two years across the North Pacific Ocean. Their long migration is modelled on an equally long gestation period of about eighteen months. During this oceanic period, the female seeks out the areas richest in food. Giant squid (*ArchItheuthis*), tuna and swordfish are her main prey. However, she is not going to refuse a good whale corpse.

For example, on 14 January 2019, Deep Blue, six metres, fifty years old, the most famous female of Guadalupe, was filmed devouring the corpse of a sperm whale, south of the island of Oahu in the Hawaiian archipelago.[18] She was enormous, much larger than usual, not because she had swallowed part of the whale, but because she was pregnant.

After gorging herself to meet the demands of her gestation, Deep Blue, like the other mothers-to-be, probably returned to the coast of Baja California, or even to the heart of the Sea of Cortez, in search of a favourable site to give birth. No one has ever shared this intimate moment. And no one knows where these famous nurseries are hidden. The only certainty is that after giving birth, Deep Blue will head back to Guadalupe, reaching it three months later.[19]

Travelling trains the young

For a long time, the movements of newborns were ignored and very rarely observed. All sorts of hypotheses were put forward: did they sink far from the surface to avoid fatal encounters with their elders? Or did they stick to coastal waters? The first study of immature fish in Australia and New Zealand showed that everyone seemed to be doing their own thing.[20] Some juveniles like coastal waters, others head north to the tropical waters of Papua New Guinea, 6,000 kilometres from their nursery. Others, on the other hand, head south into sub-Antarctic waters. The most daring one travelled almost 40,000 kilometres, the equivalent of going around the Earth, in thirty-two months!

Globetrotting sharks and drifting continents

Are the white sharks of the North Atlantic any wiser?[21] They are also great travellers and rule the entire basin, from the Gulf of Mexico to southern Greenland, via the Azores. They explore surface waters with temperatures of up to 32°C, before hunting in water at 4°C, at a depth of 1,100 metres. And there are always great explorers among them. Sharks also have their Marco Polo . . . Nukumi, a huge female, like Deep Blue in the Pacific, has embarked on an extraordinary journey along the eastern coast of the United States from Nova Scotia, before heading due east into the heart of the Atlantic, well beyond the mid-ocean ridge. It is a fantastic journey, the outcome of which no one knows, but researchers have made it available to follow live on the Internet.[22]

On the other hand, a deep mystery surrounds the great whites of the Mediterranean observed by Marcel de Serres. We know that some of them are giants. One of them, caught in 1956, 300 metres from the port of Sète, measured 5.89 metres.[23] Another, caught off Malta in 1987, was over seven metres long! Thanks to genetics, it has been discovered that the ancestors of these giants come from New Zealand, the antipodes of the Mediterranean![24] Two hypotheses can explain this astonishing DNA revelation. The first is that sharks make transoceanic journeys, switch oceans and cross the famous equatorial barrier without batting an eyelid. The second hypothesis, the one we prefer, is that the Mediterranean population is a remnant of the white shark population that ruled the world's oceans at a time when it was not divided by continents as it is today.

Indeed, 3.5 million years ago (or ten million years ago, depending on the author), the isthmus linking North and South America closed.[25] Central America separated the Atlantic and Pacific Oceans. Migrants could no longer use the narrow straits that used to connect the two oceans. But above all, this major tectonic event changed the circulation of tropical marine currents generated by the trade winds that blow from east to west. Stopped in their westward course, the currents now bounce off this continental rampart. They begin their relentless rounds in each hemisphere, as the Gulf Stream does in the North Atlantic. Marine animals are then 'centrifuged' on either side of the equator. Those from the northern hemisphere are trapped in the northern hemisphere and those from the southern hemisphere remain in the southern hemisphere, where the original home of the great whites is located. It would therefore be travellers from Australia and New Zealand, isolated in the North Atlantic after the closure of Central America, who settled in the Mediterranean. Others would be ancestors of Lady Kathy, Deep Blue and their North Pacific fellow sharks.[26]

Continental wanderings work to create species

A journey through space, a journey through time. Continental drift disperses animal populations, fragments families and quietly works to create species through geographical isolation. Fossil evidence shows that the two current species of blacknose shark (*Carcharhinus acronotus*) in the Atlantic and whitenose shark (*Nasolamia velox*) in the Pacific were one and the same 3.7 million years ago, before the closing of the Panamanian isthmus separated them.[27]

This is of course not the only cause of speciation. Minute differences in behaviour or culture are at the origin of sympatric speciation;[28] the examples of the dolphins of Shark Bay[29] or the orcas of Valdes are remarkable in this respect.[30] But the influence of major tectonic movements is often ignored, because the time scale is beyond us.

All the species that occupied the entire ocean until the Miocene (between 23 million years ago and 5.3 million years ago) without regard to an equatorial boundary were affected by the modification of surface currents due to the closure of the Panamanian isthmus. Today, apart from a few equatorial species, few marine creatures move from one hemisphere to the other.

However, the basking shark does not seem to want to follow this rule. It uses deep currents, 'hydrological tunnels', to cross the equator. Thus, this shark, which can be observed along the coast of Brittany, migrates as far as the Amazon River and even beyond the 10th parallel south, along the Brazilian coast.[31] What is the reason for these improbable migrations? As with the great white and many others: reproduction.

The great population of sharks occupies the entire ocean, from the surface to the deep sea, from the coastline to the heart of the deep blue ocean. 'Globe-swimmers' of geological times, 'globe-swimmers' of today; sharks know only one ocean. The most objective representation of the globe would emphasize the pre-eminence of the aquatic world and its inhabitants.[32] The continents are never more than drifting islands.[33]

Secret information in thick liquid

Sharks swim through the ocean as if it were their own backyard. They find their way through the seemingly landmark-free thick liquid without any problems. They migrate to their breeding grounds. They travel thousands of kilometres to the regions richest in food. How do they find their way over such distances in an environment that seems homogeneous to us?

In reality, the marine environment is incredibly heterogeneous and rich in information for those who know how to read it. Temperature, salinity, density, taste of the water and flow velocity are all clues.

The world ocean is made up of an infinite number of water masses so different that they juxtapose each other, sometimes collide, but do not mix any more than oil with water. Temperature and salinity determine their density: the colder the water (up to 4°C),[34] the denser it is. The saltier the water, the heavier it is. The combination of these two parameters largely characterizes each water mass. Some water is so dense that it literally flows through less dense water.

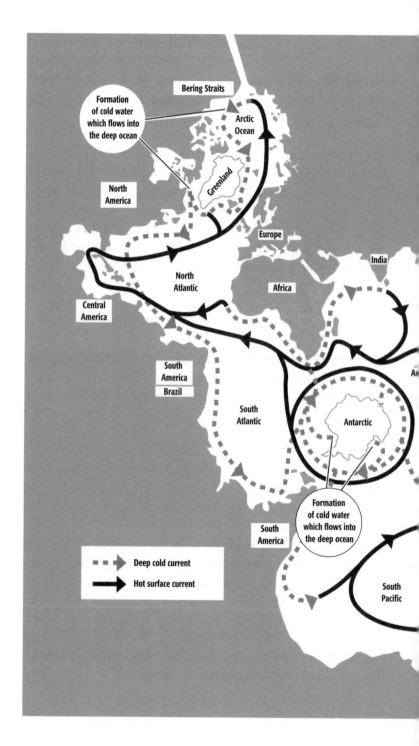

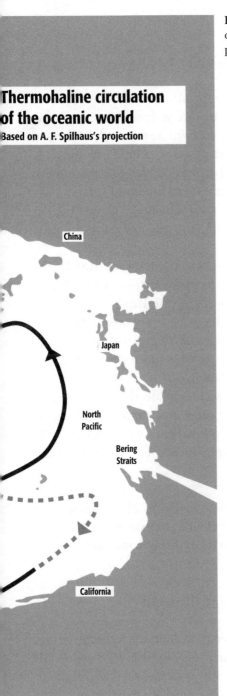

Fig. 24 Thermohaline circulation of the world ocean shown on a projection by Athelstan Spilhaus.

The oceanic 'roller coaster'

In winter, when the water temperature is below −1.9°C, ice forms on the surface of the Arctic Ocean. Freezing causes some of the salt dissolved in the sea water to be expelled. Pockets of very cold and therefore extremely dense brine form and sink. They create a deep current that flows as far as the southern hemisphere and circles the globe every thousand years.

This circulation, known as 'thermohaline',[35] carries the flavours of the Arctic region with it to the Indian and Pacific Oceans.

The departure of this Arctic water is compensated by the arrival of warm surface water from the tropical zone. This is reinforced by the regular blowing of the trade winds, which push the surface water from the tropical regions from east to west. Less regular winds – such as the Mistral – also contribute to these large movements. They push surface water out to sea and 'call in' deep water to replace it.

All these moving masses of water collide, the smells of their places of origin clashing. They generate a succession of gigantic whirlpools whose flavours are marked by the nutrient salts and plankton that develop there. The oceanic soup is infinitely heterogeneous.

Needless to say, for sharks, the boundaries of these bodies of water are markers as visible as clouds are to us in the blue sky. They are landmarks that guide them through the vastness.

A compass in the head

Even more reliable and richer than liquid references is terrestrial magnetism, which sharks use as reliably as we do with our compasses. Although a major organ for receiving magnetic fields has not yet been identified, numerous experiments have shown that all elasmobranchs perceive them very finely.[36] Not only do they detect the Earth's global magnetic field,[37] like many other migrants, but they are sensitive to magnetic anomalies caused by volcanic rocks on seamounts. Better still, they sense magnetic variations associated with tidal movements and, of course, with major ocean currents.[38] Sharks read all these signals to orient themselves, as easily as we follow road signs.[39]

Navigational error as a source of speciation

But the Earth's magnetism is not immutable. On the contrary, the drift of the magnetic poles is endless.[40] Over the last ten years or so, it has been accelerating. For example, the North Pole is drifting from Canada towards Siberia at a current rate of fifty-five kilometres per year.[41]

The drift sometimes accelerates to such an extent that the poles are reversed: the North goes south and the South goes north.[42] These reversals occur at a steady rate of once every 200,000 years, interspersed with long quiet periods. Needless to say, during periods of instability, migrants are very disturbed. Their usual points of reference lead them to unexpected regions. In addition, frequent underwater volcanic eruptions create significant local magnetic disturbances which can also mislead marine migrants.

Finally, climate change is causing major changes in the positioning of thermal fronts. All these changes in reference points mislead some of the migrants and lead the less experienced ones to breeding grounds far from their birthplaces. These lost travellers then found new colonies, and therefore new populations, the source of new species.

A brief digression on error and exaptation[43]

Navigational error forces speciation! Error, always error. Error is the engine of evolution. Before sexuality allowed for the enrichment of differences, when reproduction was limited to simple division, copying errors and DNA breakage were the only sources of difference between the parent individual and its descendants. They were the only possibilities for the change necessary for speciation. This is still the case today for the majority of living beings that do not reproduce sexually. The reproduction–copying error couplet is therefore essential to the diversification that sustains life on Earth. Life relies on the diversity of species to maintain itself when the physical environment changes. The more species there are, the more easily the failure of one species in the face of cataclysms is compensated by the emergence of another. The same is true at the population level: individual variability increases the potential and the choices that life can offer in the face of change. Difference is the master asset of life.

Life definitely does not like to be constrained by linear paths, so here is another of its twists: if it does not take advantage of an error, it diverts an organ or a behaviour from its primary function. This little adjustment (cheat?) to the instruction manual quickly proves to be a source of speciation and an asset for adaptation.

Take teeth, for example. Originally, they were used to mash food, but damselfish (*Dascyllus albisella*) use them as an instrument of sound communication. Their squeaks are amplified by the resonance chamber of their swim bladder (another little diversion) to make clicks, clacks, cracks, pulses, in short, a whole racket of a language. Very quickly, this language differs from one population to another, from one reef to another.[44] Little by little, fish of the same species no longer understand each other and therefore no longer reproduce with each other. Speciation is under way!

'*El Monstro*', the unknown from the abyss

A mistake may have written the history of the ancestors of the smalltooth sand tiger shark (*Odontaspis ferox*) that we came to formally identify, on 13 March 2001, in the waters of Malpelo Island, 500 kilometres off the Colombian coast. The holotype[45] of this shark was described in 1810 by the great naturalist from Nice Antoine Risso, from an individual caught in the Mediterranean, 10,000 kilometres away.

The sea and the black sky are mingling and trying to tear each other apart. The south-easterly wind lifts white streaks from the surface. Rocking about in our inflatable, we see the volcanic summit of Malpelo Island from time to time, like the arm of a shipwrecked man, which usually rises 300 metres above the swell. Not the most favourable conditions to go to meet *El Monstro*, the shark of the abyss, at the border of his kingdom.

El Monstro, a silhouette barely glimpsed by biologist Sandra Bessudo during an epic dive in 1998, a silhouette unknown in these parts of the Pacific Ocean. An enigma, at the dawn of the twenty-first century.

With my companions, Yves Lefèvre and Jean-Marc Bourg, I jump into the water to try to unravel the mystery. We are immediately picked up by a violent current which drives us apart in a fraction of a second. Swept up like a piece of trash, powerless, I scan the bottom which is passing by at full speed, looking for shelter. Ten metres below, the summit of a volcanic peak, a promise of respite. A mad horizontal slide to the rock of salvation. My companions, also quite shaken, join me. No doubt about it, we have left our earthly world to enter the world of the shark. We continue our dizzying descent against the volcanic rock encrusted with red algae and huge barnacles. As we approach, large green moray eels sink one after the other into their lair. Minus forty-one metres. The current fades. We pass the thermocline.[46] The water becomes cooler: 21°C. We are in the world of the beast . . .

It is there, thirty metres below. A grey silhouette gliding down a slope of shell debris. I breathe out and sink like an arrow, then inhale deeply to slow my descent and land in front of her. Depth: seventy-two metres. The crystalline water, the slightly intoxicating cold air, the size of the animal, at least four metres, and its teeth jutting at all angles . . . everything contributes to make it: *El Monstro*.

Neither Marco's lamps, which show up the scars on its sides, nor Yves' enormous movie camera frighten it: *El Monstro* is sure of itself. It does not have the lateral swaying that charac- terizes small sharks, it swims straight, without deviating from its course. It passes within a few metres. Our eyes meet. Its black eye with a white border follows me. Who am I to it? Does it see me clearly? Am I a dark mass that exhales noisy, bright bubbles? Am I the calm, steady beat of my heart? An envelope of pressure waves distorting the current? An unknown foul smell? A magnetic anomaly? What analysis can its brain, so different from ours, make of this face-to-face encounter that it has never experienced before?

It is my turn to describe the animal: it is a female with a grey fusiform body ending in a frankly heterocercal caudal fin, an impeccably round eye without a nictitating membrane, and above all a first dorsal fin in line with the pectorals. I identify it instinctively: this female has the appearance of a ferocious or smalltooth sand tiger shark.

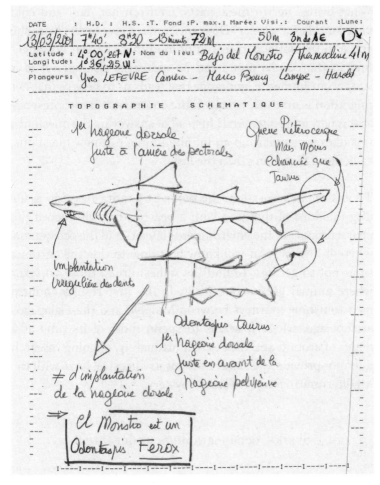

Fig. 25 First visual identification of '*El monstro*', *Odontaspis ferox*. Author's diving logbook.

But could it be the same species as *Odontaspis ferox*, identified more than two centuries ago in the Mediterranean, when there has not even been an exchange with the Atlantic for at least 3.5 million years?[47] And if so, how did it get here? Did its very distant ancestors get lost during their transoceanic migrations? Were they trapped on this side of the South Pacific?

What brings her to these parts, far from her dark and cold kingdom? Is she a loner? Does she come to give birth? Does she come to the surface in search of food that she can no longer find in the depths plundered by industrial fishing? Or is it simply our immense ignorance that leads us to believe that her migration is unusual, when, on the contrary, it is our presence that is new and ephemeral? Instead of answering the questions, this discovery opens up an infinite number of new questions, each more disturbing than the last.

To answer this question, the biologist Sandra Bessudo is fitting this lord of the deep with a beacon that will record the characteristics of the environments it visits and the ecosystems where it lingers . . . it will know best how to ride the currents so as not to get lost, to find the richest areas, those vortexes where animal life is concentrated. The first results confirm its astonishing journeys between Malpelo and the Galapagos archipelago, where it seems to spend most of its time at a depth of around 400 metres. Unfortunately, ongoing research does not provide any answers about its ancestral links with the Mediterranean *ferox*. Mystery, mystery.

Sharks, oceanographers under threat

Numerous tags have been placed on several species of sharks that visit surface waters before diving back into the deep.[48] They record the characteristics of water masses and the position of deep thermal fronts. They provide essential information for monitoring the impact of global warming on the ocean depths where life is concentrated.

In this race for knowledge, sharks[49] are very loyal allies, but we reward them very poorly. For their borderless travel across the barrier-free ocean exposes them to our deadly fishing gear. As soon as they leave their protected areas,[50] sharks are slaughtered for their fins, or caught by gear not designed for them. They are all the more exposed as their migrations take them to the high seas, lawless areas where national fleets can do no better than the poachers of misery recruited by international traffickers.

Will sharks disappear before they are even understood?

8

Fading Silhouettes

11 June 2014. The Tunisian port of Zarzis is blue and white. An armada of wooden ships of all sizes, huddled together, raises a tangle of masts to the sky, bearing the red national flag stamped with the crescent and the star. An incredible maritime caravanserai permeated by the sticky smell of guts rising in the morning heat. The docks disappear under mountains of octopus pots. Sailors sit on the ground, repairing their nets. It is an adventure to make your way to the next longliner coming in. It arrives pretty late, having come from afar. It has just returned from Libyan waters, which until now have been forbidden and virtually untouched.[1] The port is in turmoil. What we see coming out of the boat's hold is staggering: forty or so fifty-kilo groupers, a dozen two-metre grey sharks weighing sixty kilos, a small sharpnose sevengill and dozens of monstrous red scorpionfish, at least fifty centimetres long. All caught in one night!

I find this 'miraculous' catch deeply upsetting. These corpses with their eyes forever dulled are those of particularly venerable patriarchs. Some must be over fifty years old. A generation that will never be replaced, neither in our lifetime nor in that of

Fig. 26 Unloading of grey sharks at Zarzis, Tunisia, on 14 June 2014.

our children. The rhythm of our exploitation is so intense that the fish no longer have time to age.

Spectacular fishing never to be repeated. You would have to wait fifty years, without any harvesting, to find so many old groupers and scorpionfish, treasures in terms of reproduction, because their fecundity is much higher than that of the young. For sharks, it is more serious, because their fecundity is naturally pitiful.[2] In the Mediterranean, it is not just the giant old individuals that have disappeared forever, but entire species for which not even a juvenile can be found.

A richness lost, forever forgotten. This astonishing harvest also highlights our ecological amnesia.[3] It shows that the generous and overabundant Mediterranean described by ancient naturalists was not a fantasy:

> Every year we see innumerable armies of emigrating fish arriving on our coasts . . . Anchovies, sardines and mackerel appear first . . . They are followed by tuna and sharks, which always come last. So we see the mackerels eating the sardines, just as

the tuna eat the mackerels, the tuna themselves being devoured in turn by the sharks, which pursue them with such relentlessness . . . that they allow themselves to be washed ashore rather than suffer the cruel death that awaits them . . . Among the sharks, the white shark, *Squalus carcharias*, is the most common on the coast of Provence . . . One of its usual companions is the glaucous shark (blue shark), *Squalus glaucus* . . . It can reach up to five metres! The longnose shark, *Squalus cornubicus*, [porbeagle shark, *Lamna nasus*] exceeds 300 kilos, as does the smooth hammerhead shark, *Sphyrna zygaena*, known in Provence as the *gat* or the cat[4] . . .[5]

The naturalist Marcel de Serres wrote about these sharks in his 1840 reference book. Where have they gone? To answer this question, the film director Stéphane Granzotto and I conducted a long investigation in the fishing ports around the Mediterranean.[6] France, Italy, Algeria . . . the encounters are endless but the answers are similar: 'You should have been here thirty years ago. All that is left is Tunisia, and the Gulf of Gabes . . . The most important shark breeding area in the Mediterranean. Even the great white comes here to give birth.[7]

In search of the last grey sharks

13 June 2014, 33° 29′ north latitude, 11° 34′ east longitude. The Mediterranean, off the Gulf of Gabes, is being buffeted by a violent north wind. *Kais II*, Kamel's fishing boat, which has just set its gillnet, is dragging its anchor. And the small raft on which we have embarked for a week is being hammered by legions of whitecaps constantly breaking. In the distance, the red buoy that marks where the net starts regularly disappears under the surface, indicating the strong current. Diving along beside the shifting and lethal mesh will require our full attention.

But we hop into the murky water with cameraman René Heuzey, hoping to film grey sharks attracted by fish caught in the net. We are dreaming of emerging from the water all smiling as witnesses to their presence. Hope turns to despair: entangled in the mesh, three drowned green turtles, a dozen corpses of sharks barely fifty centimetres long devoured by sea snails and a spiny butterfly ray (*Gymnura altavela*), a critically endangered species. No large sharks, either alive or dead . . . Same result after six more dives and a week of effort, at the end of which Zarzis's best fisherman returns empty-handed . . . '*Inshallah!*'

No, not '*Inshallah*'. Overfishing, which killed off the last Mediterranean sharks. The grey shark, which used to reproduce by the hundreds close to the coast in the Gulf of Gabes, made the fishermen of Zarzis rich. They said the grey shark was inexhaustible, but now it is gone. In ten years (2007–17), despite the opening of new fishing sites in Libyan waters, catches have fallen by 43 per cent.[8] Not only are the numbers of sharks inexorably decreasing, but the size of the individuals is often smaller than the size of sexual maturity! The boats are going out of business. Our friend Kamel has become a sponge fisherman.

Northern Mediterranean sharks have been gone for even longer. I saw the fabulous 5.89-metre white shark, caught in 1956, 300 metres from the port of Sète, whose stuffed remains are on display in the Lausanne museum. I met Jean Licciardi, the last fisherman from Sète to have caught a great white, a few hundred yards off the coast, in 1991. As for other species, after a peak in catches (25,000 tonnes) in 1985, the die was cast at the end of the 1990s. Ninety-seven per cent of the great blue and mako shark population had disappeared![9] Commercial fishing is not the only factor. Recreational fishing is definitely the nail in the coffin. Scientists from the International Union

Fig. 27 Blue sharks caught during a 'sport' fishing competition in Port-Camargue, 1980.

for Conservation of Nature (IUCN) point out how catastrophic this recreational fishing is, as amateurs, too often unaware of protection regulations, go after the last survivors that professional fishing has left behind.[10] Smaller species, which are often more prolific, are not spared either.

Diving on a carpet of sharks

The memory of a dive comes back to me, 8 August 1971, off La Couronne (Marseille). At a depth of forty-five metres, the water is not really clear, as it often is at this time of year. Millions of microscopic crustacean debris, sea urchin larvae and starfish, agglomerated globigerinas, form fluffy clusters that remain suspended throughout the dense liquidity. My plunge down into the midst of these organic snowflakes is vertiginous, but the arrival on the bottom definitely feels wobbly. Nitrogen narcosis? Not at minus forty-five metres! Yet the bottom is slipping away. The bottom is moving. The bottom is entangling me. The bottom is alive. I am stuck in a carpet of 'sea dogs'.[11] Brief notes in my logbook authenticate this memory, which does seem exaggerated. With time, memory fades, but my quick jottings on paper, as if nothing was going to change, remind us of a bygone Mediterranean. 5 August 1972, still off La Couronne, minus forty meters: fifteen dogfish gathered and huge conger eels under the coral vault. 5 August 1971, lobsters everywhere under a coral ceiling, huge and very numerous scorpionfish, conger eels, lobsters . . .

All over the world, the same story

Different times, different places, the same abundance. The amazing accounts of the oceanographer-photographer Anita Conti, who sailed for thirty-five years (1934–70) on fishing boats throughout the Atlantic, seem to be the stuff of legend today.[12] Some of her photos are simply breathtaking. In one of them, soberly titled 'Return from shark fishing, Senegal, 1945', there are seven huge tiger sharks, among them a man is struggling to pull another shark onto the beach.[13] In the background, a group of crouching men pay no attention to the scene: just another day fishing, everyday life . . .

Tiger sharks were so abundant on the West African coast that the oil extracted from their enormous livers was exported and the shops were full of their frightening jaws. The Cape Verde archipelago was particularly renowned for the size and number of these large predators. When, in November 2001, I went there with photographer Fred Bassemayousse to write a piece on the 'famous Cape Verde tigers', we had taken all the necessary precautions in case we were bothered by any big sharks who meant business. We did a lot of diving with the best local guides, but we did not see any sharks. And yet, plenty of impressive jaws were on display in the beach shops. 'They're from Africa. There are no sharks here', we were told.

If Hemingway were to write *The Old Man and the Sea* today, what shark would his hero be defending his legendary sword-fish against? There are no more sharks in the Caribbean, no more sharks in the Gulf of Mexico. The history of the region's fisheries is instructive: between 1940 and 1979, a small arti-sanal shark fishery survived to satisfy local demand. Despite this, it hurt the shark population. Natural abundance is quickly forgotten: even if the species are not endangered, they are irre-versibly deprived of their large older individuals. Between 1980 and 1998, intoxicated by the international marketing of shark meat and fins, the fishery was industrialized. The sharks were decimated; the fishery closed at the beginning of the twenty-first century.[14]

All around the world, the same bloodbath has occurred: from Australia, where catches of great white, tiger and hammerhead sharks have collapsed by 74 to 92 per cent,[15] to Brazil, where the smalltail shark (*Carcharhinus porosus*), which used to be so abundant that no one paid any more attention to it than to sardines, is now critically endangered,[16] and the high seas, where the whitetip shark is gone.[17]

According to the latest International Union for Conservation of Nature report (2021), the tragedy is accelerating; the number of threatened species has doubled since the first global estimate was made only seven years ago: 391 species of sharks and rays, or more than a third of chondrichthyan species, are in danger of extinction. Worse, of the 1,093 species assessed, 99.6 per cent are threatened by intentional and incidental overexploitation.[18]

Fins of shame

Fishing is by far the most important cause of declining shark numbers, as sharks are either targeted directly for their meat and fins or caught in fishing gear that is not intended for them.

Thirty-eight million sharks are slaughtered each year for their fins. The large pelagic species (blue, silky, hammerhead, mako) pay the highest price.[19] It is an appalling slaughter. The sharks arrive in single file, hooked on longlines several dozen kilometres long. When they are hauled aboard, their fins are removed and they are immediately thrown back into the sea, alive but mutilated. The fishermen do not bother with the large bodies, which are worth 10,000 times less per kilogramme than the dried fins, which can be sold for US$1,000 per kilogramme. By comparison, shrimp in the same markets are sold for US$6 per kilo.[20]

In addition to these estimates, which are based on official sales, there is also the black market, the extent of which has only been discovered by chance. On 24 September 2021, Colombian authorities found nearly 3,500 shark fins on their way to Hong Kong, even though shark fishing is banned in Colombia. A year earlier, on 7 May 2020, customs in Hong Kong, the world's second-largest shark fin market after the Chinese port of Guangzhou, seized twenty-six tonnes of dried fins taken from 38,500 sharks. Worse still, a DNA investigation of the Singaporean market revealed that among the sixteen species of sharks and rays slaughtered, two species are strictly prohibited from trade and six others are in danger of extinction.[21]

And these tasteless fins are not even used to feed people; they give soup a 'texture'.

The hideous beauty of shark squalane creams

While the Asian and American markets have the highest demand for fins, the European market, a major consumer of beauty creams, is not to be outdone. One in five creams uses squalane, a shark liver oil derivative that helps the cream penetrate the skin. 'It is estimated that three million deep-sea sharks are killed each year specifically to meet the international demand for squalane. For some, nearly 95 per cent of the population has been decimated',[22] explains Claire Nouvian, director of the Bloom Association, which has conducted a major investigation into the murky market for shark products.[23] The cosmetics market is constantly growing, increasing the demand for shark liver oil. It mobilizes more and more research laboratories to find new extraction methods,[24] new anti-ageing creams[25] and new drugs for lung cancer.[26] And never mind that deep-sea sharks are in danger of extinction. So much the worse if they could be spared because the equally moisturizing plant-based substitute exists, but seems too expensive to provide substantial benefits in the ultra-short term.

Driving on shark oil!

To top it all off, some researchers are planning to make 'biofuel' from shark liver oil![27] When you create the supply, you must then satisfy the demand. If we open this Pandora's box, it will swallow up deep-sea sharks and all the others.

The management of misery

All species are threatened, even those that are not currently in danger of extinction and whose exploitation is 'managed'. Because 'not in danger of extinction' does not mean the 'fullness' and 'abundance' of even half a century ago; far from it. It is precisely when they are not yet under threat that it is important to take radical measures to protect species. Because we know that these animals with a low renewal capacity[28] cannot withstand the pressure of industrial fishing. We must not wait for the alarm to go off. Then it is too late, always too late. This concerns the species of small sharks, falsely sold under the name of 'salmonette'. 'Salmonette' is not a small salmon, but a school shark, or a spiny dogfish (*Scyliorhinus* sp., not endangered), or a houndshark (*Mustelus asterias*), whose name is not mentioned. They are served 'masked in sauce' and are used in school canteens and catering facilities.[29]

Fishery management avoids 'total collapse'. But it relies on auctions, which may be deceptively unstable, as has been shown for the school shark. While managers gave it the green light, its population collapsed by 88 per cent and its global IUCN status quickly changed from 'vulnerable' in 2014 to 'critically endangered' in 2020.[30] And now it is too late to apologize: 'We didn't anticipate . . .'.

Unfortunately, bone-free 'salmonette' meat, which meets the 'safety standards' for mass catering (e.g., nursing homes, school canteens), is becoming increasingly attractive. As soon as one species become extinct, the others will be under increasing pressure, which is unsustainability on a global scale. More seriously, this 'management of misery', which is based on an inordinate demand, will never allow the return of the schools of shark that were the delight of our dives just fifty years ago. As for the large, old individuals that characterize healthy eco-systems, their time, like that of the dinosaurs, seems to be over forever. Our children will have to make do with faded photos or giant stuffed corpses in museums.

Shark flesh, deposits for heavy metals

Like all marine creatures, sharks and rays are affected by anthropogenic pollution. The most dreaded toxins (pesticides, PCBs, lead, cadmium, mercury)[31] are concentrated in their flesh,[32] to the point where, in some cases, it is unfit for human consumption.[33] However, it is difficult to determine the impact of these pollutants on sharks' lives.

A poisoned shark does not complain

Apart from the immediate and fatal consequences of cata-strophic pollution, the impact of chronic pollution on living organisms is very difficult to identify and quantify. Non-lethal doses can cause significant physiological disturbances over time.[34] And sick sharks don't come complaining. So how can we demonstrate a lower resistance to disease and temperature variations? How can we link a drop in fecundity, embryonic malformations or neonatal mortality to a combination of low concentrations of chemical agents? We are unable, even in humans, to measure the synergy of the effects of different toxic agents. What we do know is that the heavy metals, PCBs and other biocides we send into the marine environment are not conducive to the development of life.

The appalling consequences of eating bluefin tuna on the health of fishermen in the port of Minamata, Japan, have alerted us to the deadly danger of methylmercury in humans.[35] The impact appears to be just as severe on fish, which initially see their equilibrium disturbed and their opercular (gill) movements accelerate before dying. As for other heavy metals, such as cadmium and lead, they cause significant alterations to the liver and gills, muscle degeneration and skin necrosis.[36]

In addition to these heavy metals, macro- and microplastics have contaminated the entire food web. Here again, the same analyses confirm that all sharks are affected, without exception:[37] large and small, deep-sea[38] or open-water, planktivores[39] and mammalian predators.[40] Here again, the same long-term concerns remain, with no answers.

Divers who keep track of known individuals confirm these fears. For example, in the Shark Reef Marine Reserve in Viti Levu, Fiji, researchers monitored a bull shark injured by a fishing line in its jaw for seven consecutive years. The wound gradually developed into a monstrous cancerous tumour.[41]

What about climate change?

What can be said about global warming? Here again, how can we understand it, since the changes impact the ocean as a whole, on all ecological scales: global, regional and local? It modifies the density of water, and therefore the major marine currents. It increases the temperature of surface waters, disrupting planktonic life and, consequently, all food webs. It is impossible to predict how different species of sharks will adapt to it, benefit or suffer from it. However, the impact of local warming on certain species can be measured in the laboratory. For example, the embryonic development of the bamboo shark (*Chiloscyllium punctatum*) is profoundly disrupted when its eggs are kept in water at 28°C instead of 24°C.[42] Seasonal increases of this magnitude in surface waters are becoming increasingly common around the world.

Some migratory species (lemon shark,[43] bull shark), whose movements are induced by temperature change, will probably take advantage of this warming to conquer new territories that were previously unfavourable to them. Always very opportunistic, they are moving northwards along the Atlantic coast of the United States. They are now breeding as far north as North Carolina, where they were only occasionally seen in the past.[44] These colonizations are the result of particularly exploratory individuals, such as Frizzia, the bull shark who gave us some great encounters in Playa del Carmen in Mexico in 2019 and 2021. Her transmitter tag has already signalled her forays into North Carolina seven months after she left Playa.[45]

Change of territory, new threats

These forays into areas where swimmers are not accustomed to the presence of sharks increase the probability of accidents and, as a result, the implementation of policies to eliminate these newcomers, considered to be intruders. Hunting these sharks to secure tourist areas can have dramatic consequences on species with low fecundity such as the bull,[46] the white or even the tiger. This is why the tiger shark population in south-eastern Australia has been completely wiped out.[47]

Every day, humans colonize more coastline to treat it as their playground. The sharks, on the other hand, are making the most of their changing territory. Confrontation is becoming more and more frequent. How can such confrontations, sometimes dramatic, be transformed into more peaceful encounters? Is the elimination of sharks for the sake of unrestricted leisure by the sea necessary? Is it legitimate?

9

The Confrontation

Réunion Island, February 1981. 'Don't go diving at Cape La Houssaye; it's infested with bull sharks and tigers.' I don't much like the advice of my Réunionese friends. Black sky, anthracite basalt, white foam. The westerly swell violently hits the rocky outcrop standing up to the Indian Ocean. Cape La Houssaye was not a pleasant place to be that morning. Down under the surface and into the first few metres of muddy sediment, the bottom is visible through the milky water. An understated scene. The violent westerly storms did not allow any reefs to flourish. Only a few purple algae and robustly shaped corals encrust the rough volcanic rock. Apart from the school of common bluestripe snapper (*Lutjanus kasmira*) which illuminate the dark bottom without really brightening it up, there is little life. A solitary Chinese trumpetfish (*Aulostomus chinensis*) shelters behind the fan of a gorgonian swaying in the current. A loggerhead sea turtle (*Caretta caretta*) takes off with a beat of its flippers. I had a good look in the deeper ocean, hoping for the silhouette of the big predators I had been expecting, but nothing. No sharks.

It was probably not the day.

Are there days when sharks are easily seen in Réunion? No. It is even extremely rare to see them while diving.[1] Sharks are neither very abundant nor very curious. Although many species frequent the island, including the great white shark,[2] Réunion Island is certainly not a shark hot spot, unlike other parts of the world where you are almost certain to see them on every dive.

Bull shark as villain

And yet, thirty years later, in November 2013, I dived again in Réunionese waters in search of the bull sharks suspected of having caused five fatal accidents in three years. This appalling series of tragedies provoked a wave of terror and irrational elimination fishing (culling). But above all, it triggered an incredible controversy surrounding the National Marine Nature Reserve,[3] which was accused of having 'deregulated the ecosystem' and of having encouraged a shark population explosion by setting aside reserves as resting and breeding areas.[4] With no inventory, and no counts before or after these accidents, there is nothing to indicate that sharks are more abundant than before. But there is a time for media attention and tears, and a time for scientific research.[5] Researchers from the Institut de recherche et de développement (IRD), who are hard-pressed to provide an immediate explanation for the tragedies, were being taken to task. As for the representatives of the Marine Reserve, they are held responsible because 'together with the scientists/ecologists, they have played at being sorcerers' apprentices by transforming an aquatic kindergarten into a marine reserve'.[6] It was at their request that I went to the site, after a first unsuccessful attempt at mediation by Patrice Bureau, president of our association Longitude 181.[7] In order to get around policy issues, I proposed that everyone could go and see the situation for themselves. With the help of diving clubs,[8] we invited everyone, ordinary citizens, representatives of the administration, surfers and swimmers, to dive with us to see the abundance of sharks directly underwater and to base their opinion on the facts. The results of these dives were edifying: we did not see any sharks.

But it is too late. No one wants to believe the researchers' facts or the underwater observations any more. They are demanding revenge. They want an eye for an eye. Irrationality has inflamed them. No one seems to ask if 'punitive' fishing is relevant or if the punishment should be of an 'exemplary' nature. As if the threat of 'capital punishment' were going to be a deterrent among the wider shark population. No doubt about that.

The only response the authorities are proposing to the grieving families is setting an example by killing sharks. They increase the number of hunts. There are scenes of hysteria and jubilation when a captured shark is slaughtered on the beach. Whether the shark was involved in an accident or not is irrelevant. All are guilty! On the other hand, the sharks do not protest and are of little consequence compared to the seaside economy and surf schools.

In addition to punitive 'post-attack' fishing, there is also culling. Amateur fishermen are encouraged to eradicate the threat by offering them gift cards in supermarkets for each shark caught.[9] Citizen participation in the slaughter is praised.[10] To eradicate sharks, longlines baited with tuna are set in swimming areas and even in the heart of the reserve's reinforced protection zone.[11] The state did not set any limits on the number of sharks caught or the duration of the fishing. The pressure must be maintained, whatever the cost . . . More than €11 million has been spent to date.[12]

Bull shark as saviour

Playa del Carmen, Mexico, 14 January 2021. Putting on my diving suit, I'm hoping they will be there. Everyone here in Mexico hopes they will come back every year. Everyone wants them to stay as long as possible . . . What are they? About sixty female bull sharks. Yes, they are the same species of shark as on Réunion Island: *Carcharhinus leucas*. But here in Playa del Carmen, one of Mexico's most popular tourist areas, not only are the bull sharks pampered, but fishermen are paid to take their hooks and nets elsewhere. And if an inattentive shark is inadvertently caught, the fisherman is compensated at the price of a kilo of meat to release it . . . This is how Playa del Carmen became one of the best places in the world for a face-to-face encounter with the bull.

And its reputation goes unchallenged.

Barely in the water, I can make out their dark, paunchy silhouettes twenty-five metres below, moving in arabesques on the light sandy bottom. The pectoral fins spread out on either side of the massive head and support the rear of the streamlined and supple body. The bull shark has that rare elegance that combines power and elasticity. There are five females, four of which have the rounded bellies of mothers who will give birth in the next few days. They swim slowly, sure of their strength, so close to the bottom that their pectorals regularly raise small clouds of sand. They are tracing enigmatic paths on this large, uniformly flat plateau. Beside me, a dozen other divers who have never been in contact with such diving companions. Some of them are apprehensive about this first confrontation, others dreaded it so much that they didn't dare tell their parents that they were going to dive with bulls.

One of them approaches. Perfectly stabilized in the liquid mass by its pectoral and dorsal fins, it looks like a missile with a squadron of small trevally. The opposing current seems to boost her propulsion. Under the square muzzle, the slightly open mouth is visible. And then the anthracite eyes that harpoon me and never let go. Three metres. I hold my breath and flatten myself further against the sand, but it is not enough. The big female does not want to give more at this stage. She bends her course and presents me with her right flank. I recognize her: it's Jenny. Her first two anastomosing gill slits form a sort of X that is easily identifiable, so much so that since the scientist Mauricio Hoyos drew up her identity card in 2009, each of her visits to Playa del Carmen has been recorded. For at least eleven years now, this beauty has been delighting observers.

'Delight' is a weak word to describe the neophytes' exhilaration. Back on the boat, they were excited at having overcome their quite understandable worries, and told stories of encounters that were very different from what they had imagined. This encounter changed them forever. Never again will they calculate sharks in tonnes. Each one will be considered an individual, a non-human person who deserves consideration. Never again will they witness a killing without thinking about the living individual, without questioning the usefulness and legitimacy of the massacre. This consideration cannot be learnt from books, it must be imbibed with all one's senses, with one's heart, with one's gaze locked into the 'Other's'.

Better alive than dead

A massive tourist business has grown around shark observation. The few dozen sharks that spend the winter in Playa del Carmen bring in several million dollars each year. This is far more than what the same sharks sold at the fish market would have earned. Better still, these sharks have welded together a divided community. The Saving Our Sharks Association[13] has brought together competing dive centres and fishermen to work together to preserve the bull sharks and secure an annual income. Each diver who comes to see the sharks pays a fee to the association, which in turn gives part of the profits to the fishermen so that they do not set their lines in the area.

Mexico is not the only place in the world where divers flock to rub shoulders with sharks. The Bahamas attract thousands of divers to Tiger Beach to see tiger sharks and great hammerheads (*Sphyrna mokarran*) as well as bulls. South Africa has an abundance of white, bull and sandbar sharks. Better still, Fiji, whose economy is largely based on bull shark diving, has had to preserve the integrity of its coral reefs in order to support the presence of these large predators. Sharks are thus the keystone of the entire reef ecosystem and the human economy. They are both a source of income and, indirectly, protectors of the reef ecosystem that humans no longer exploit to maintain their presence. On all tourist reefs, the diversity of marine species is much higher and the individuals are much larger and older than on traditionally exploited reefs.[14]

Why so much hate, why so much love?

Why are the same sharks, so popular in one region, symbols of terror elsewhere? Are the accidents, always horrifying, always unacceptable, more numerous there? What lessons can be learned from these tragedies?

First, it is necessary to describe the accidents as accurately as possible. To date, the International Shark Attack File (ISAF) is the only organization in the world that collects and scientifically analyses accidents involving sharks.[15] Its scientists rightly distinguish between provoked accidents (when humans intentionally interact with sharks) and unprovoked accidents, when humans did not intend to interfere with sharks. The annual global average of unprovoked accidents has been almost constant over the past thirty years at eighty, most of which result in minor injuries. Unfortunately, and again the number is almost invariable, about a dozen are fatal.

Lightning worse than sharks

By way of comparison, if you splash around in the sea along the US coastline, you have a one in 264.1 million chance of being killed by a shark, but a one in 3.5 million chance of drowning. The marine environment itself is dangerous, infinitely more so than sharks. Another example, again in the United States, is that in the last forty years, twenty-six people have been killed by sharks, while at the same time 1,970 people, seventy-five times as many, have been killed by lightning. Even jellyfish kill at least five times as many people as sharks (probably many more, as only accidents involving tourists are taken into account).[16] However, beach activities expose their users to unfortunate encounters very unevenly. Ninety-seven per cent of accidents involve surface activities: 61 per cent involve surfers and bodysurfers[17] and 36 per cent involve swimmers. The remaining 3 per cent involved anglers and underwater hunters. Very few divers were involved.[18]

Improbable diving accidents

However, the exceptional happened on 25 September 2018 to the 'Under the Pole' expedition team, during the ascent from a deep dive on the Bora Bora reef, in French Polynesia. The decompression stop was to be endless: two and a half hours before returning to the surface. At a depth of eighty metres, visibility is exceptional. Ghislain Bardout, the expedition leader, films his chief diver, Julien Leblond, being nonchalantly pulled along the vertiginous drop-off by his underwater scooter.[19] A grey shark, like the thousands that exist in Polynesia, is on the prowl. A grey shark like the team sees every day without paying attention, a shark that is not considered potentially dangerous, approaches Julien. Its behaviour is unusual: pectoral fins quite lowered, swimming slow and jerky. It circles around Julien. Suddenly, it lashes out, bites him violently on the head and swims away. It will never come back. Despite the cloud of blood that spreads, no other shark approaches the injured diver, whose ordeal has just begun because his diving apparatus has been destroyed. He cannot go straight to the surface or he will die of decompression sickness. Exceptional diver, exceptional composure. Julien calmly goes over his emergency equipment. Ghislain is there. The team surrounds him for 150 minutes during which anything can happen. Finally, everyone is safe and sound on the boat. Ten stitches for Julien; a deep psychological trauma for the whole team.

During the next day's dive, the slight apprehension is partly overcome thanks to the objective analysis of the event. Because the incident, which was quite exceptional, is perhaps not so inexplicable. The signs of extreme nervousness of the shark could (should?) have alerted the divers. The contortions of the body, with the pectoral fins clearly lowered, are obvious signs. They were well described at the end of the 1970s by Donald Nelson and Richard Johnson, who pointed out that the use of electric underwater propulsion particularly excites sharks.[20] In hindsight, Ghislain Bardout himself emphasized that

> although the incident was unforeseeable, it is not completely inexplicable. This behavior of the grey shark is known as a territorial intimidation swim. It was probably trying to say 'Get out of here'. At the time, it was difficult to interpret. The use of rebreathers,[21] which allow us close contact with the animals, does not work in our favour in this case. The bubbles from classic suits frighten the animals. And, after this incident, we did not hesitate to release a few bubbles from our rescue tanks when a shark seemed a little too 'curious'. Caution should

therefore be exercised with a shark, just as with an unfamiliar dog. While we should obviously not demonize sharks, we should also remember that they are wild animals, with whom we must interact with caution and humility.[22]

Biologist Johann Mourier experienced a similar incident, fortunately without consequences. While being towed by an underwater scooter in the waters of Fakarava Atoll, he was charged in the same way by a grey shark. He had the presence of mind to stop his machine, causing the shark to stop its contortions instantly. These contortions were resumed as soon as the scooter was in motion.[23] Could the vibrations of the scooter and the electric motor be the cause of the animal's agitation?

The charge of the great white

I have experienced similar situations many times, but the shark did not go through with its charge.

More specifically, I remember a great white shark charge in Guadalupe on 12 November 2006, when we were shooting the film *Oceans*. We had been in the water for almost an hour, cameraman David Reichert and myself. The visibility is remarkable, over thirty metres, especially with the midday sun beaming right down into the water. Since the beginning of the dive, Ugly, Lady Mystery and Trigger, three large females, have been circling around us without real conviction. Suddenly, a new female, which we had never seen before, enters the dance, without us being particularly disturbed. Usually, the more sharks there are, the more they watch each other, and the less interested they are in us. However, the newcomer is nervous: her jerky swimming alternates between sudden accelerations and phases of calm. Suddenly, she lowers her pectoral fins, arches her back and opens her mouth wide, then gathers

Fig. 28 The author, while diving in Guadalupe, is filmed by René Heuzey sketching the attitudes of white sharks.

herself up with arched back and charges me. She repeats this succession of contortions several times, accompanying it with a generous defecation that betrays her own anxiety. Seasoned by several dozen hours of diving with great whites, I do not react to her demonstration of intimidation. And, at about five metres from me, she curves off and disappears. The painless warning was not be repeated.

Demonstrative intimidation charge by a great white shark on the author

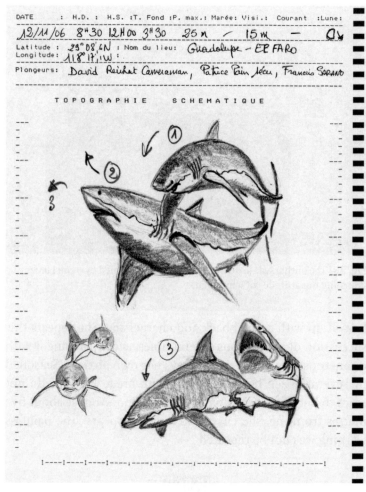

DATE   : H.D. : H.S. :T. Fond :P. max.: Marée: Visi.:   Courant :Lune:

12/11/06 8"30 12H00 3"30   25m   ⌐   15m   —   ⊙⋎

Latitude : 29°08,6N : Nom du lieu: Guadelupe - EL FARO
Longitude: 118°17,1W :

Plongeurs: David Reinhat Cameraman, Patrice Tain Adeu, Francois Sarano

TOPOGRAPHIE   SCHEMATIQUE

Fig. 29 Contortions of the great white shark during her charge. Author's diving log.

Sharks attack to eat

But sometimes there is no warning. Diving instructor Céline Lefebvre had a dramatic experience of this, even though she thought there was no risk when scuba diving.

Bourail, New Caledonia, 31 December 2008.[24] It is the beginning of the dive. Céline calmly descends with her group along the reef. Nothing to report, except the breathtaking beauty of the coral cliffs. At a depth of thirty-five metres, without having seen anything coming, the young diver feels a terrible bite. A large tiger shark has bitten her deeply on the leg.

A month later, Céline is still in hospital. She has had countless operations. In bed, with a smile on her face, she talks about her accident with humour. 'I couldn't get closer to a tiger shark than that!' she says, repeating that sharks must be protected. 'A wild animal at home in an ocean whose diversity and richness we must preserve.'

Making the unacceptable intelligible?

How can we explain accidents? It seems impossible. Every accident is a special case. The only thing that can be said is that no rules can be drawn from them! Because the number of accidents is, fortunately, insufficient to be statistically analysed, as the places, the hydrological conditions, the circumstances and the species involved are so varied. And even more so, because each shark has its own personality which leads it to react in its own way.

Eric Clua, a shark specialist, places particular emphasis on this individual factor. He shows that only a very few individuals, particularly exploratory and daring, will attempt a test bite, whereas most of their fellow sharks will not do so under the same circumstances.[25]

At most, factors can be identified that favour a dramatic confrontation or a peaceful encounter. Accidents occur more often in the evening and when the sea is rough. But here again, it comes down to case by case, species by species, as lifestyles are so specific. For example, bull sharks come very close to the coast, even in the mouths of rivers, during the breeding season. They prowl coastal areas where rain has washed away odorous organic waste. Thus the cocktail of 'winter season + bad weather + rain + gully mouth + sunset' considerably increases the probability of a bad encounter.

The only lesson that can be learned is that it is better to avoid swimming in such circumstances. It is important to learn about the habits of the sharks that inhabit the area where you are going to practise a nautical activity to avoid finding yourself in a dangerous situation.

One confrontation, two actors

Let us not forget that there are two actors in a confrontation: human and shark.

The encounter is not perceived in the same way by either of them. While the shark immediately perceives the intrusion into its domain, most beachgoers are unaware of what is happening below the surface. They neither see nor feel the presence of sharks, even though the sharks may pass by indifferently a few metres away. Videos taken from a drone that can 'see' through the surface show great white sharks swimming nonchalantly among carefree and unaware swimmers and bodysurfers.[26]

This lack of awareness greatly amplifies the shock when the shark appears.

The encounter is then experienced as a 'surprise attack', a travesty. A whole cascade of responses flows from this feeling. And the response of the authorities is based on the public's perception of accidents. For what is important for the authorities is not so much to solve the problem of the rare accidents as to restore public confidence.[27] In the end, it is less a shark affair than a social psychology affair. The response varies not according to the number of sharks and accidents, but according to society's idea of individual responsibility/freedom or, on the contrary, its disempowerment in favour of the welfare state. Another important factor is media coverage, which amplifies the phenomenon, thus orienting the authorities' response even more.

Finally, each actor, human or shark, has his or her own personality which, when the environment is particularly unfavourable, may or may not lead to tragedy.

And while sharks have not changed fundamentally in recent years, humans have profoundly altered their understanding of the marine environment with the explosion in beach tourism since the beginning of the twentieth century.

Sailors and consumers of seaside leisure activities

Our pioneers were sailors, connoisseurs and lovers of the environment as a whole. They embraced the ocean in which they regularly immersed themselves. They accepted the rules, the dangers and the sharks. These seafarers were joined, then overwhelmed, by 'beachgoers'. Hundreds of millions of people who play, swim, surf, dive and occupy shark territory in all seasons thanks to improved equipment.

These city dwellers who spend a few days by the sea are often unaware of the rules of the marine ecosystem and do not want to know them. They go to sea with the carefree attitude of those who visit a safe amusement park. They have paid for a restful holiday and nothing should disturb them. They rely on the authorities to remedy their inconsistencies.

This shift in attitude is highlighted in Christopher Neff's study of the changing public and official responses to a series of similar accidents off Sydney between 1929 and 2009. In 1929, not only did the government not blame sharks, they blamed irresponsible bathers going swimming in the evening when sharks were hunting. To avoid accidents, they insisted on education and called for individual responsibility and a change in behaviour.[28]

From 1935, a turnaround took place. Now it seemed to be a given that the coastline was a zone reserved for bathing and recreation. The culprits were no longer bathers, but recreational fishing and sewage that attracted sharks close to the shore. Under public pressure, the government set nets to catch sharks that came close to the shore. It also installed lookouts to restore the confidence of bathers. These measures were taken for a period of two years and were considered experimental. Seventy-five years later, in 2009, the nets are still there and the government is blaming . . . sharks 'for swimming near the beaches and jumping into the nets'!

In surfing, the obsession with sporting competition is sometimes combined with the certainty that beach recreation is a given. Some athletes no longer consider the ocean as a living environment but as a wave generator. They consume waves as if they were in a gym, refusing to submit to natural constraints. And above all, to the presence of sharks on a territory they consider their own.

Avoiding accidents?

How can we at least reduce the probability of unfortunate encounters, which is obviously what everyone wants, although there is no consensus on the methods?

The simplest method is to isolate bathers from sharks by separating them with huge nets along the most frequented beaches. This method, used along the coast of KwaZulu-Natal in South Africa, is proving successful: only two minor injuries were reported between 1980 and 2011. However, while bathers can enjoy 300 kilometres of beach to swim safely, dozens of dolphins and turtles, whales, birds and at least 400 sharks die, entangled in their nets, every year.[29] This 'collateral' damage to many protected species raises the question: can our recreational activities in the wilderness be at the expense of its inhabitants?

Culling only causes collateral damage

Moreover, these nets cannot be set along steep coasts, which are highly exposed to swells and storms. So, in response to the anxiety of tourists, the authorities financed massive fishing operations. This culling policy, which began in 1959 in Hawaii, the home of surfing, quickly showed its limits: hundreds of thousands of dollars were spent and 4,660 tiger sharks were killed, without any noticeable reduction in the number of accidents over twenty-five years.[30] In light of this, the authorities have abandoned culling and have made shark fishermen face their responsibilities. So much so that a bill for the full protection of all sharks in Hawaiian waters was introduced in 2019.

In Australia, each state has its own policy. While the state of New South Wales pursues a culling programme, in Queensland the Humane Society International has obtained a ban on shark fishing on the Great Barrier Reef.

The main drawback of these culling programmes is that the shark responsible for the accident is rarely caught. In the end, these fisheries only cause collateral damage! On Réunion Island, for example, there have been considerable blunders. As of 31 August 2021, after seven years of baiting, the official balance sheet[31] shows that 170 bull sharks have been killed, none of which could be linked to an accident. Worse, 387 tiger sharks and 735 other sharks (hammerheads, corals, nurse sharks) and fish were also captured by the deadly bait. And it doesn't matter that these sharks, often juveniles, were not involved in the accidents . . .

Culling policy assumes that 'every shark is a potential culprit'. It is reminiscent of the fable of the wolf and the lamb: 'If it is not you, then it is your brother or one of your own' who is disturbing my clear water.

This disastrous outcome does not fundamentally change the risk of accidents. It does not absolve surfers from the necessary precautions that must be taken in all cases, as recommended by all surf instructors in Réunion since 2006, after two fatal accidents: 'The sharks are at home . . . We must respect the sea and its rules . . . We must stop surfing after 4pm and especially in winter . . . not surfing when conditions are not favourable . . . we must surf responsibly.' They called for 'a revision of the basic rules for safe surfing'.[32]

Destruction without limits

The other problem with this policy of safety through eradication is that it is limitless, because the demand for security is absolute and limitless. First, to satisfy the demand for 'revenge', the state decided to eradicate – at random – a fixed number of sharks. But this number is not based on any ecological reality, because there has never been a prior estimate of shark abundance. This estimate is only made a posteriori, by studying the evolution of catches. Yet fishermen bait sharks in order to catch them. A terrible vicious circle. Fishing attracts sharks to the coast that would otherwise never have come, increasing the probability of encounters with swimmers and justifying the continuation of the elimination programme!

This is especially the case for small oceanic islands visited by migratory sharks that travel across the ocean. The steep underwater slopes of these volcanic islands force fishing gear to be set very close to the coastal swimming areas. The bait therefore attracts the sharks to the very place where we would not want them to come! In Réunion Island, for example, the 387 tiger sharks that usually frequent the open sea were attracted by bait placed a few hundred metres from the coast. A high point for a politics of protection.

Eric Clua, who believes that the personality of sharks – rather than their profusion – plays a major role in accidents, stresses the need to look for individuals that are willing to bite a human for predatory purposes. He points to the possibility of genetically identifying[33] these bold individuals in order to selectively remove them and avoid the carnage of indiscriminate culling.[34] Rather than a random slaughter, this case-by-case preventive culling could be understood and accepted.[35] But this strategy is difficult to implement. Above all, it is not currently psychologically acceptable to those who have made the elimination of sharks their life's work.[36] It looks like defeat in a confrontation with the 'savage', and a negotiation with a horde of assassins.

Need for caution

The exorbitant cost of these safety measures pushes politicians to outbid each other in the media to emphasize that they have 'the situation well in hand'. So, reassured and relieved of responsibility, sea-users inevitably relax their attention. Sometimes they even transgress the basic safety rules laid down by the authorities whose protection they need. No matter how many sharks they take, they will never exempt sea-users from learning and respecting the rules of caution in a wild ecosystem.

So why not start there?

Other solutions that are less catastrophic for ocean populations have been proposed: aerial surveillance, underwater surveillance, personal protection.[37] But it is always a question of securing a territory that we humans have decided to appropriate.

It is perhaps on this point that the defenders and the sworn opponents of sharks differ most radically.

The misunderstanding

Does the sea belong to the world of leisure? Isn't that the root of the problem? The misunderstanding stems from the fact that some people consider sharks, bulls in particular, to be invaders of a coastal area dedicated to water games. The book *Comprendre la crise requins à la Réunion* (Understanding the Shark Crisis on Réunion Island), which brings together the testimonies of various stakeholders, states: 'The balance of the ecosystem depends on the *reappropriation* of marine space by humans.'[38] Without human pressure, there can be no ecological balance: 'The real problem is that on Réunion Island, sharks are hardly fished any more, as they are in other hot spots around the world, which are affected by this problem.'[39] 'Fishing is necessary to maintain the biological balance'. But above all: 'Sharks must know the limits of their territory and must learn to stay in their place',[40] far from the coastal zone and the marine reserve, which belong to water sports.

Is the ocean a wilderness or a theme park?

Is it legitimate to demand the elimination of sharks so that we humans can indulge our recreational whims in complete ignorance and without constraints?

In some regions, we are even entitled to ask the question: is it reasonable to develop a nautical activity in an unsuitable place since it will have to be made safe at great expense by destroying the ecosystem in which it is practised? Should we, because surfing exists and is practised in Polynesia or Biarritz, develop it everywhere in the world? What would we say if we installed ski slopes in the paths of avalanches on the pretext that skiing is practised elsewhere? Or if we equipped climbing routes on cliffs with friable rock because climbing is fashionable? Perhaps surfing is not the right water activity during the winter period in Réunion.

But the priority for economic growth, combined with the idea that the customer is always right, stops people asking the prior question of the relevance of the activity being developed. As a result, security has to be installed without discussion.

How can we change our perspective?

How can we change our perspective if not by going to meet the person we want to understand in his or her home? How can we economize on immersion in the world of the other? Because this immersion allows us to know.

Far beyond knowledge, 'knowing' (co-naître) is a feeling, a life experience, a 'living with' that nourishes all the senses, that forges intuition beyond analysis. The encounter allows us to grasp the behaviour, to decipher the body language of the shark. It allows us to know the 'codes' to avoid misunderstandings, so that the encounters remain beautiful.

The diver quickly understands how discontinuous vibrations and surface impacts stimulate sharks. It is not surprising that a bodysurfer, or a surfer who lies on his board and paddles with his hands, arouses their interest. It's a clear call: he's luring the sharks by simulating an injured fish! And while few sharks are mistaken, there are some overly excited, very bold and hungry individuals who do not hesitate to bite. I remember in 1988, when *Calypso* was in Papua New Guinea, we were amazed by a fisherman from the Hermit Islands who waved a coconut rattle on the surface of the water as a shark call. After about ten minutes, a shark came alongside the canoe and the fisherman lassoed it.[41]

How far can we go to change opinions?

Not everyone has the chance to dance freely with a great white shark. It is even rare to see sharks up close when diving, as they maintain a safe distance. How can we break the cycle of ignorance and indifference? Should we go so far as to offer those who do not know how to dive the opportunity to discover sharks under artificial conditions?

Some diving clubs organize events during which instructors, wearing chain mail, feed sharks by hand. In a row, equipped with scuba gear, onlookers get a close-up view of wild sharks gobbling up barely thawed dead fish . . .

This practice raises two questions. Firstly, what behavioural changes does regular feeding induce? Secondly, what image of nature does the observer get from it?

In response to the first question, a six-year study of bull shark feeding sites in Fiji shows that the behavioural changes are minor and don't last long.[42] The shark is an opportunist and remains an opportunist: you bring it food, it takes it. If you do not feed it, it does very well on its own. As for its seasonal migratory movements, they remain unchanged, food or not.

From Charybdis to Scylla

Feeding, on the other hand, profoundly alters instantaneous behaviour. Not only does the abundance of food concentrate the sharks abnormally, which compete for food, but the sharks, totally uninhibited, go straight through a group of divers without paying them any attention, even colliding with them. The sharks expose themselves. Excited by the food, they are caught up in a frenzy comparable to that observed when they gather around a whale carcass. They accept that their fellow sharks and divers are intruding on their personal space, reacting at the last moment. The same shark that would normally have turned away from a diver at a distance of three metres comes within one metre.

This extreme proximity has its advantages. It removes the fear of the divers and turns the most cautious of observers into a shark defender because, when their eyes meet, the bond that is established is unbreakable. Objective achieved, you may say.

Yes, but what is the objective? To increase the number of shark protectors? Then it's achieved! But if the objective is to reintegrate oneself as a 'shard of life' among 'Other', non-human persons, then it is hopelessly lost.

For the show reinforces the belief that nature is there to satisfy our cravings for 'everything, right now', light years away from the reality of the wild where hours of patient waiting, often to no avail, put us back into the cycle of time and allow us to enjoy rarity. Even with free sharks in the wild, the show maintains the idea that humans are above other living beings and have the right to manage, exploit, preserve and enjoy a nature that is external to them and which they can dispose of at will. This staging perpetuates the 'culture/nature' dualism that the encounter with unpredictable wildlife is supposed to erase.

Indifference, the deadly enemy

Today, however, there is an urgent need for action, as sharks are disappearing amidst a general indifference. There is an immediate need for a chorus of voices calling for the protection of these unloved ones before it is definitely too late. The few studies that have been conducted on the benefits of 'shark diving'[43] show that divers lose their preconceptions about sharks and become more committed to their protection.[44]

No one is indifferent. Even those who have no real fascination with sharks are transformed. The encounter has a long-term impact. It strengthens the conviction of already concerned observers and creates new ambassadors for the shark cause.[45]

I ask myself the same questions. And I remember smiles lighting up the peaceful faces of companions who had observed our dives with the great whites in Guadalupe in 2013. They had stayed in the cage, admiring the sharks passing by a few metres away. Despite the bars, they had felt the magnificence of the animal in their whole being. They had felt something that cannot be expressed in words, something that cannot be quantified but is experienced, and that instantly reminds us that we are all part of a living fabric.

Their happiness and humility spoke volumes about the awe-inspiring shock they had just experienced. In their eyes was the certainty that these splendid aristocrats had their place in this world and that it was unacceptable to imagine their disappearance. Even if artificial, these encounters had broken the *indifference* that is the mortal enemy of our relationship with 'Others'.

Who speaks for the sharks?

'I hope this dreadful bite can be used to save sharks!' Céline Lefebvre speaks calmly and with hindsight about the terrible injury inflicted on her by the four-metre tiger shark. 'The accident even strengthened my passion for the underwater world and for these majestic animals in particular. They are so badly treated, even though they play a vital role in the balance of the ocean. Their survival depends on ours.'[46] Three weeks after the accident, although she is still intubated in Noumea hospital, Céline talks about devoting part of her life to telling her story in schools. She wants to 'reach out to young people who, all too often, do not know the sea, even when they live on an island'. She wants to make people love sharks.

'Respect the sharks that injured us'

The members of the Fin for a Fin surfers' association have the same conviction and are committed to respecting sharks, the very ones that have injured them.[47] They are in favour of peaceful cohabitation: 'We surfers know the risks when we get into the water. However, when there is an attack, it is not us, it is others who demand revenge and organize shark massacres in the name of those who have been bitten … without their consent! We surfers don't want any more massacres in our name.' This community has many champions, such as Mike Coots, Paul de Gelder, Rob McDowell, Claire Bevilacqua, Blair Kimber. On the fin of their surfboards is written: 'If my life's taken don't take theirs'. They say: 'We want to protect the shark that bit us. Because if the ocean is our playground, it is first and foremost the world of sharks. Our community wants to stop the vicious cycle of hysteria and revenge and create a cycle of coexistence.'

Fortunately, one does not have to have been bitten to call for peaceful coexistence. On the other hand, most of those who are committed to protecting sharks – always a difficult and sometimes dangerous struggle – have spent time with them. Often, they have known them since they were young. They have shared their world peacefully. The empathy they feel towards them is so strong that they themselves feel attacked and feel that the killing of sharks is an intolerable injustice. They fight to defend their 'family'.

'A dead shark is a wound in my flesh' (Sandra Bessudo)

Sandra Bessudo is in charge of one of the largest shark sanctuaries, the Malpelo Marine Reserve off the coast of Colombia.[48,49] She braves the very real threats to her life without batting an eyelid. Because protecting sharks is her life's work. 'It's hard to explain why I fight for sharks. Every time I see a dead shark, I feel it as a personal wound, as if it were in my flesh. It hurts me.' She first put on her mask and dived at the age of four:

> I was captivated by an angelfish. A little later, off the coast of Malpelo, I was overwhelmed by the abundance of barracudas and trevally, by the multitude of sharks. My appreciation of this changed me forever. When you are immersed in their midst, when you share moments with them, you begin to know them. Never again can you see sharks as food stocks. Never again can you describe them in tons and dollars. Even today, for the Colombian authorities, sharks are not part of biodiversity, they are just fishery resources to be exploited and managed. This

is because managers and fishermen have not gone to see the sharks where they live, underwater. We need to bring nature to life for children, as I was lucky enough to do thanks to my parents.[50]

Sandra does not hesitate to take part in the boarding of pirate ships fishing in the sanctuary. She also does not hesitate to go to the ports to meet the fishermen, despite the threats. She wants to convince people. And in order to persuade the harder cases, she offers them the chance to dive with her in Malpelo's waters. 'But not to fish, of course.' So, Apolinario Celorio, one of the fiercest opponents of the fishing ban in the Malpelo sanctuary, agreed to dive with her, even though he had never put his head under water. He came out dazzled. Hoping that 'this moment will remain forever in [his] dreams', he decided to go and share his wonderment with his fishing friends so that they would respect the marine reserve.

In January 2021, Sandra won another battle. She won a ban on shark fishing in all Colombian waters, as well as a ban on the marketing and export of sharks. But some fishermen violently rejected these measures and openly repeated their threats.

Sandra even believes that sharks can be used for diplomacy, to bring people together. A case in point: 'her' hammerheads, highly migratory sharks that travel from Malpelo in Colombia to Cocos Island in Costa Rica, via Coiba Island in Panama, and stay in the Ecuadorian Galapagos archipelago, have managed to get the governments of these four countries to the table. Together, they are working on the creation of a marine reserve corridor where sharks can swim in peace.

'Being with sharks without frightening them'
(Steven Surina)

Founder of the Shark Education association, Steven Surina has been diving with sharks since the age of six.[51] He spent his childhood in contact with them, travelling through the Red Sea on his parents' dive boat. Every day he learns how to approach them 'without frightening them'. Having discovered the right distance so as not to violate the shark's personal space, he can then get as close as possible to understand its behaviour. With this invaluable experience, he encourages divers, experienced or novice, to forget their preconceptions and to discover for themselves this sphere of intimacy specific to each shark. And above all, he wants to share the emotions that sharks transmit to him at each encounter:

> I have never experienced or felt as much intensity, respect, empathy, peace and love as after a dive with sharks. They leave such a deep imprint on us that it affects us for the rest of our lives. It is impossible to accept that they might die out. Future generations should have the chance to experience this at least once, to stop demonizing sharks forever. It is this paradox that escapes me, that of finding total serenity in the company of an animal that embodies fear itself . . .[52]

Rob Stewart, Sea Shepherd, Longitude 181

The same is true for Rob Stewart, whose film *Sharkwater*, denouncing the appalling slaughter of sharks for their fins, shocked millions of viewers.[53] As a child, at the age of nine, he came across his first shark, and he quickly became involved in their protection. Paul Watson, founder of Sea Shepherd, was forever marked by the sight of a sperm whale harpooned to death.[54] And all those, less famous, such as Didier Dérand, in Réunion Island, and many other representatives of associations who, like him, fight in the field, sometimes risking their lives, to save those they consider their brothers.

What brings us together in our association for the defence of marine life, Longitude 181,[55] what motivates our commitment, is this powerful empathy towards our wild *coloca-Terres*.[56] This fraternity, forged through contact with them, must be helping to break down fear and indifference.

Otherwise, how can we feel like brothers? For those who do not go swimming under the skin of the ocean, the animals that live there seem as alien as Martians. Why should they care about beings they don't know and who don't exist for them? They have to make a considerable effort, appeal to reason, to save abstract beings that do not concern them.

French Polynesian sharks saved by divers

In 2003, when our association Longitude 181 became involved in preserving sharks in French Polynesia, which were threatened by massive exploitation for their fins, it relied on a group of Polynesian divers who were in daily contact with sharks. Because they had lived with these wild creatures, they were the only ones getting angry. Although Polynesians culturally respect sharks and do not fish them, the authorities, like the population, remained indifferent to the development of this new fishery. Some even saw it as moving economic development towards the Chinese world. It took four years of daily mobilization, gathering 40,000 physical signatures (petitions were very rare and did not yet exist on the Internet), and every week bombarding the mailboxes of Oscar Temaru's government, from the headquarters of the Longitude 181 association. Finally, the Minister of Sustainable Development, Georges Handerson, issued a first decree in 2006, and then in 2007 a law that definitively banned the fishing and marketing of sharks.[57] It is thanks to those who have felt the gaze of sharks that French Polynesia is now their largest sanctuary in the world. It is thanks to those who mobilized that, ten years later, Laurent Ballesta's team was able to film *700 Sharks in the Night.*[58]

Following this success story, many other countries have taken conservation measures that have ensured the return of sharks. In time, even such fragile species can repopulate the ocean.[59]

But you have to want it to happen!

So what are the reasonable arguments that would force us to save sharks? What can we do to go even further? To hope for Harmony?

10

Reconciliation

So why should sharks be saved? They attack swimmers. Their presence disrupts the development of seaside economies. They consume fish that could feed those in need. After all, dinosaurs are certainly extinct, and it is part of natural selection to see species disappear. Sharks just don't belong on our Earth any more . . .

Let's try to go through the reasons making a case for their conservation. Biological, psychological and philosophical reasons.

The death of sharks unravels the web of life

Each living being, each species, has a place in the ecosystem, defined by the set of relationships that the individual weaves with other living beings and the physical environment. The more complex the being, the more numerous and complex the relationships it weaves. Consequently, the more its disappearance has consequences for other species and the ecosystem. The resilience of life is based on redundancy, on the formidable entanglements that unite all creatures. The greater the uniformity, the more fragile life becomes; the less it can rely on the diversity of its assets to offer solutions in the face of change.

It is both the interweaving of species and the variety of their capacities that make life resilient. Sharks are among the ligaments that hold the web together. And, make no mistake, humans are only one of the threads (*fils*).[1] Each species that disappears unravels the magnificent tapestry of life. Despite our ingenuity, and our formidable technical resources, if we were to be the last, we would become aware of our solitude and precariousness very late.

What is worse, the species we are wiping out are precisely the most complex ones, those that live a long time, those that have a late sexual maturity and often a low fecundity. By doing so, we are opening the door to all the rapidly renewing species that are not always the ones we cherish the most: jellyfish, worms, rats. These very ubiquitous, very tolerant species always take advantage of the resources left available by those that disappear. It takes hundreds of millions of years of random successive mutations to provide the elegance of a Lady Mystery. Should she disappear, nothing will ever replace her.

Necessary unpredictability of life in the wild

Human dominance of the world is narrowing our universe. What would a world be like if it were populated only by species that were bred, monitored, fattened up, dependent, enslaved. Where is the magic? What would a world be like if we suppressed all the creatures that annoy us? Do we want an Earth where representatives of every species, once free, are preserved for posterity in giant aquariums and zoos, pathetic and derisory Noah's arks? Such dominance imprisons us in our own security. Our free space closes in. What kinds of choices, responsibilities and freedoms does a theme park offer? In our desire to escape the unpredictability of sharks and any others that escape us, we build the walls that enclose us. On the other hand, the unpredictability of creatures that do not respond to our rules broadens our horizon; it is a window on dreams, surprises and the marvellous. It is a school for encounters, where we patiently learn the right distance to tame the 'Others', those who are irreducible.

What kind of world do we want to offer our children? Monocultures, aquaculture farms, protein production sites? I would be unhappy to have to tell my children and grandchildren: 'When I was on *Calypso*, I was able to appreciate the last great wild animals. They gave me immense joy, moments of fulfilment. But I didn't know how to pass these things on to you.' Worse: 'I knew. Everyone knew. Scientists had sounded the alarm. All the governments of the world had signed the Rio appeal in 1992. We were fed with figures that measured, step by step, biodiversity decline. We had statistics that gave us the clearest picture . . . and I did nothing to give you sharks in the wild.'

We can escape into virtual worlds that are truer than life, whose 'reset' possibilities suggest lives can be lost or taken with impunity, but it makes us light-headed. The shark, the last great symbol of wild life, is there to remind us of the world to which we really belong. A sensory, sensitive and sensual world, a world of flesh. Sharks re-establish empathic links with 'Others' that the virtual universe is unravelling. They reinsert us into the great cycle of life and death, clarifying those existential questions we vainly try to escape from by losing ourselves in fake controlled universes.

Sharks eliminated for convenience: who will be next?

If we accept the elimination of sharks today, not for reasons of survival but on a whim, to reduce our irrational fears, to carry out recreational activities totally free of responsibilities, where will we stop? How many other species will we eliminate for our own convenience? Where do we draw the line? How many other *coloca-Terre* non-humans and . . . humans?[2]

Some people, forgetting that we are part of the living fabric, irrevocably linked to all other living beings, filter species for utility. They thus separate the wheat from the chaff, the good fish from the sharks, the useful from the harmful. If we use the filter of profitability to judge what should or should not be conserved, who will we retain, given that standards vary so much with time, territory, tradition, culture and philosophy? Am I useful? Will I be preserved? On the other hand, if we know how to make room for the cumbersome and the insignificant, we will know how to make room for each of us, human and non-human, with our differences and our singularities.

What would a world without sharks be like?

But . . . is the question relevant? We are already in a world without sharks![3] So why are we trying so hard to save their last remaining representatives? And, reasonably, the answer is not obvious. Perhaps it is because they are marine animals far from our everyday world, so far away that we feel neither their absence nor the richness of their presence. Marine creatures simply do not exist in our frame of reference as earthlings.[4,5] So let's ask ourselves: what would a world be like without gorillas, tigers or elephants? What would a world be like without foxes, finches and frogs? What would a world be like without Mozart and Leonardo da Vinci? The answer is painful: the world would not be very different and at the same time it would be *fundamentally* different.

In-different

We humans have wiped out or pushed to the brink of extinction so many species that, unfortunately, we are able to say, in hindsight, that the world without them is not very different. Holed up in our cities, 'above ground', lost in our new virtual worlds, we have not seen the ecological upheavals that these disappearances have brought about. We do not even complain about the extreme poverty of the monocultures we call 'nature'! We have exterminated species even in our thoughts. They were once chaffinches, wagtails, goldfinches, warblers, chiffchaffs . . . now there is only 'bird'. They were gudgeon, minnow, grayling, rudd, tench . . ., now there is only 'fish'.[6] They were millions, all different, they are now the 'nature-sun' setting in which to spend a holiday. They are now only images in children's books. We are Mowgli lost in megacities, orphaned by Bagheera and Shere Khan. We are Romain Gary, whose letter is returned to sender with the note 'no elephant at this address'.[7] We are the Little Prince without the fox. All we know of the dodo, the Tasmanian tiger and the great penguin are their effigies on the back of coins. Soon our children will be asking themselves about where the insane imagination came from that created the extraordinary bestiary in the animated film *The Lion King*.[8]

Unfortunately, we already live without them! Just as we could very well live without the paintings in the Chauvet cave and the masterpieces in the Louvre. Just as we are totally indifferent to the fate of the prehistoric frescoes of Messak Settafet or the temple of Apollonia subjected to the ravages of war in Libya.[9]

Fundamentally different!

However, we feel that the disappearance of a species, like that of a work of art, impoverishes us irreparably. Beyond the ecological changes that this disappearance entails, we understand that this loss contributes to the involution of the world. It goes against the formidable movement that has been at work since the dawn of life on Earth, 3.8 billion years ago: the 'Great Natural Diversification'. Diversification that human genius has extended through extensive cultural and artistic diversification. Is diversity not one of the few criteria that 'objectifies' the sense of the history of life? Don't diversification and complexification due to associations between species indicate what we might call the progress of life? Is it not diversity that makes the planet rich? The loss of a species through our own fault is a regression in history. Is not each disappearance attributable to us an indignity that denies what we aspire to be: human?

Indeed, is it not through this relationship with the wild world that we best demonstrate what defines us as human? No other species asks itself the question of respect for other creatures and their preservation. This question is unique to us. Is it not, therefore, this respect that builds us, that makes our identity? Is it not by the yardstick of respect for 'Others' that we measure our humanity?

It is up to us to make peace with shark non-sense

The shark has no reason to exist. There is no reasonable reason to preserve it, to spare it, except that, precisely, this decision to save it, matured freely in our consciousness, projects us into another dimension. A dimension that can no longer be reduced to the sole physico-chemical dimension of the universe, an ethical dimension which transcends its origin and which, in this world of meaninglessness, offers meaning.[10]

Isn't the 'New Alliance' with the wild a consent to merge with it, to think of ourselves as brothers in alterity, even in death? This consent makes us different from all other creatures and, in revealing meaning, it offers peace.

It is therefore up to us to change the paradigm, to move from a logic of annexation to an understanding of living together in a world to which we all belong, human and non-human.

Because, in the end, it is a question of finding a diplomacy – as the philosopher Baptiste Morizot suggests[11] – that will allow us to live in peace, while, every day, we humans colonize more territory belonging to others, without caring about the rules of their ecosystems, despising their existence, their demands and their codes. In so doing, we multiply the risks of dramatic confrontations. And the accident, suddenly, leaves us helpless. The unbearable, unacceptable horror of this inexplicably sends us back to our arrogance, reinforcing suffering and igniting hatred.

It is up to us to find a diplomatic way to get around the logic of 'for sharks' versus 'against sharks'. It is up to us to blend into the global ecosystem and to have consideration for everyone, human and non-human. As Baptiste Morizot rightly suggests, we need to adopt the politics of regard, a diplomatic tool for escaping the dualistic logics (nature/culture, sacred/profane) imposed by modern naturalism.[12]

In order to deal with the most pressing problems, some of my environmentalist friends reinforce this moral dualism by sacralizing certain species. By banning access to nature reserves, essential meeting places, they extract man from the world, reinforcing the idea that there is humankind and nature.

In fact, humans are nature. But to really understand it, we must do an 'apprenticeship'. And this is done on the ground, looking eye to eye, not just in books.

Take the risk of being clumsy!

To those who say that there are too many of us to go out and meet wildlife, which is currently confined to a few tiny reserves, I say: let's multiply the reserves. Better still, let's take the wild-life out of the wild! If there is such a demand for nature, it must be satisfied. Instead of reducing the number of privileged people who share moments with free and untamed creatures, we must increase the possibilities of encounters tenfold by encouraging the return of the wild everywhere!

To those who say that humans are too destructive and aggressive to let them come into contact with animals, I reply: you cannot learn to meet them on some Internet 'tutorial'. You have to take the risk of the awkwardness of the first face-to-face encounters, the risk of making mistakes. One can only experience the right distance from the Other by trial and error.[13] The singularity of each person can only be detected through contact. And this discovery of *singularity* means being considerate to every living being. For those who come into contact with sharks, there are not just 536 species, there are tens of millions of individuals, all unique.

Can we not imagine 'masters of respect' who can lead us to discover this singularity, just as there are teachers of commerce or technology? It is up to us to give priority to learning about encounters and, as Albert Jacquard suggested, to inscribe on the pediment of each of our schools: 'Here, we teach the art of encounters'.[14]

But above all, let us not give up learning just because there are risks. Let us not give up on becoming *Sapiens*.

Lady Mystery

Perhaps Lady Mystery, the great white shark, holds a key to our lives?

November 2006. Filming of the film *Oceans*,[15] Guadalupe, Pacific Ocean. Lat. 29° 08′ north; long. 118° 17′ west.

Already thirteen dives and more than twenty hours trying in vain to get close to the great whites, although there are plenty of them. Twenty hours with them, awkwardly groping to overcome the wary face-to-face encounter that characterizes every approach. My noisy bubbles and my awkward enthusiasm probably had a lot to do with it. But how do I know at what subtle moment to move towards the shark without frightening it? How do you find the right distance without experimenting?

Yesterday, Trigger, the impulsive female, left the imprint of her teeth on my drawing tablet and, in trying to get closer, I scared her away for the rest of the day. The harmony wasn't really there . . .

Patient and tenacious, cameraman Didier Noirot allowed himself to be approached to within a metre to film Lady Kathy's inquisitive black eye. She had the confidence not to take off . . . This is a real improvement. Is she the one who has changed? No, it's Didier who knew how to be more discreet.

Day after day, we find sharks that do not know us and have never met a human in their world. Day after day, the contacts are more harmonious. Day after day, we learn how to behave. We feel what it takes to be with them. My ignorance and clumsiness fade away in favour of a subtle, indefinable tranquillity, calm, lowered gaze, waiting, arms pressed to the body . . . that I cannot really describe. An intuitive, animal feeling that I would be hard pressed to put into words, figures or equations. A whole attitude full of discretion, envy, benevolence. Every day I feel a little more ready.

And on this 12th day of November, this time with David Reichert, I am carried away from the boat by the current. The sun is hazy, the dark blue water is full of plankton. Lady Mystery comes up to us, the biggest of all the females we have seen: five and a half metres, one and a half tons. Her pectoral fins are horizontal, a sign of calm. Does her size give her confidence? No swaying of the body, she moves straight ahead. Nothing seems to be able to stop her course. Her implacable assurance communicates something. I feel a soothing serenity. I slide along her right flank. We swim together, shoulder to fin. We brush against each other, respectful. A few centimetres separate us. But in this minute of eternity, we are one body. Harmony. No artifice between us. An authentic encounter, without calculation, one of those encounters that give us the profound joy of communing with life. A rare gift offered by an untamed, free animal, and therefore incredibly precious. A totally unnecessary and indispensable gift.

Five years later, on 29 November 2011, same place, different film, with my friends Stéphane Granzotto and René Heuzey.[16]

The water is not very clear. In the sun, ten metres above, like a guardian angel, Aldo Ferrucci, my 'rebreather master', is ready to intervene if the respirator on my back starts to play tricks on me. I can abandon myself to the 'benevolent' solitude of the immensity of the ocean.

And there she is. I don't recognize her immediately. I'm 'doing my job': on my drawing board, I write down the depth, the time of arrival, the estimated size. I sketch her silhouette to put the characteristic marks that will allow me to identify her. But her massive elegance makes me lose track of my notes.

I know how derisory this information is. It is so reductive that it is a travesty of the creature I want to know. I can estimate her weight, determine her sex, describe the line of her coat, draw her attitudes, the movement of her fins. But how can I describe the perfect fluidity, the wild elegance, the deep emotion that she conveys and that moves me forever?

At five metres, I recognize her, it's Lady Mystery. She accepts me again. We swim, once again, shoulder to flank. The distance between us is not measured in centimetres, it is measured in mutual trust and, for my part, in respectful consideration.

She moves away. I want to keep her here. Soon I can only make out the broad movement of her giant fin plunging into the blue night. Yet the more she fades, the closer I feel to her: it is precisely because she does not respond to my whims, because she escapes me, because she is free, that Lady Mystery is indispensable to the world.

I am at peace. A deep serenity overwhelms all my senses. A moment of fulfilment.

Lady Mystery calls to me. She forces me to question this shared moment, to meditate on my relationship with the wild. It is because everything seems to separate us that this moment of peace forces me to reflect on what unites us. Lady Mystery plunges me into a vertiginous questioning of my place at the heart of life. She forces me to pay attention to all the other human and non-human *coloca-Terres*: warbler, finch, ant, oak. Paying attention to others in nature takes me away from the dizzying diversions of everyday life and forces me to ponder existence.

Beyond our singularities lies consideration

But how did she, Lady Mystery, experience this moment? Did she even experience it as a special moment? Probably not. I don't know. Nobody will ever know. I don't think we'll ever understand sharks. Too much separates us. Our singularities drive us irreparably apart.

Fig. 30 Lady Mystery and the author, shoulder to shoulder.

November 2011, another appointment with
Lady Mystery.

Fig. 31 Lady Mystery carries the author on her pectoral fin.

Do we have to rely on intermediaries to break through the impossibility of understanding others, as Baptiste Morizot so brilliantly suggests?[17]

Is it not enough, quite simply, to strip ourselves of our shell of preconceptions, as I did with Lady Mystery? To surrender oneself, authentically, in confidence? To lay oneself bare, to listen? Is it not enough to learn to receive? I don't think we need to rely on therianthropes, interpreters or shamans to allow us to communicate with the world. Lady Mystery tells me that it is enough to give it consideration and kindness. It is enough to say 'hello' thoughtfully, genuinely, sincerely.

Then, for a moment, although Lady Mystery and I could not understand each other, we were at peace. This harmony, the very epitome of the plenitude that puts us in tune with the world. And what I believe is that this peace depends only on me! It depends on my hope to find the right distance. It depends on my freely exercised will to understand and accept. This willingness is unique to us humans. It is our originality, among all the other originalities of the non-humans who make up the abundant tree of life.

Beyond our singularities, which distance us to the point of impossibility of understanding each other, is not the will to understand stronger, more necessary, than understanding itself? Is it not the prerogative of humans to have this will? Doesn't it build our humanity?

It is up to us humans to make peace. That is my hope, and it lies entirely in this faith in our formidable *difference.*

Notes

Introduction: Giving the 'Voiceless' a Voice

1 Steven Spielberg, dir., *Jaws*, Universal Pictures, 1975.
2 Jacques Perrin and Jacques Cluzaud, dir., *Océans*, Galatée Films, 2010, and in English, *Oceans*, Disneynature, 2010.
3 Romain Gary, 'Letter to an Elephant', *Life Magazine*, 22 December 1967.
4 *Umwelt* is defined by the biologist Jakob von Uexküll in his book *A Foray into the Worlds of Animals and Humans: With A Theory of Meaning*, trans. Joseph D. O'Neil, Minneapolis: University of Minnesota Press, 2010. It is not only the environment as our senses allow us to experience it, but also the dialectical construction between our perceptions and our actions acting on the environment. Each living creature is therefore constantly constructing its own world, its *Umwelt*, in all its singularity.
5 Baptiste Morizot, *On the Animal Trail*, Cambridge: Polity, 2021.

Chapter 1: A Matter of Misunderstanding: From Pliny to Disney

1 Jacques Cousteau and Frédéric Dumas, *The Silent World: A Story of Undersea Discovery and Adventure*, London: Hamish Hamilton, 1953.

2 Jules Verne, *Twenty Thousand Leagues under the Sea*, Project Gutenberg eBook, 1994.

3 Awareness of changes in the world's ecosystem definitely moved beyond specialist circles to the general public with the front page of *Time* magazine on 2 January 1989, which chose 'Endangered Earth' as its Person of the Year. That year, the date for exceeding the planet's renewable resources was 11 October. In 2021, it was 29 July. See https://www.overshootday.org/

4 'The State of World Fisheries and Aquaculture', Food and Agriculture Organization of the United Nations, 2020. https://www.fao.org/publications/sofia/2022/en/

5 *Longimanus* is in reference to *Carcharhinus longimanus*, an oceanic shark with disproportionately long white-tipped fins, hence the comparison with long arms.

6 Jacques-Yves Cousteau and Philippe Cousteau, *The Shark: Splendid Savage of the Sea*, trans. F. Price, New York: Arrowood Press, 1987, p. 32.

7 *The Jacques Cousteau Odyssey*, TV series, 1977.

8 Cousteau and Cousteau, *The Shark*, p. 5.

9 Shark Sider/Shark Mythology: www.sharksider.com/shark-mythology

10 Aristotle, *History of Animals*, Books 7–10, ed. and trans. D.M. Balme, Cambridge, MA: Harvard University Press, 1991, pp. 445–7.

11 Pliny, *Natural History III*, Books 8–11, trans. H. Rackham, Cambridge, MA: Harvard University Press, 1983, pp. 214–15.

12 The seas mentioned by Pliny are in the Mediterranean: the Sea of Sirte, the coasts of Tunisia and Libya, and those of the Aegean Sea.

13 Pliny, *Natural History*, pp. 264–6.

14 Anacharsis was a Scythian philosopher of the 2nd century BCE.

15 Jean-Marie Kowalski, 'Les marins et la mort: actualité d'un mythe', *La Revue maritime*, No. 492, 2011, p. 90.

16 Greek philosopher and scholar of the fifth century BCE.

17 Greek philosopher and scholar of the third century BCE who lived in Cyrene, in modern Libya.

18 Monique de La Roncière and Michel Mollat du Jourdin, *Les Portulans: Cartes marines du XIIIe au XVIIe siècle*, Fribourg: Office du Livre S.A., 1984.

19 Pierre Belon, *L'Histoire naturelle des estranges poissons marins*, Paris: Hachette Livre BNF, 2012 [1551].

20 Adriaen Coenen, *The Whale Book: Whales and Other Marine Animals as Described by Adriaen Coenen in 1584*, eds. Florike Egmond and Peter Mason, London: Reaktion Books, 2003.

21 Guillaume Rondelet, 'Libri De Piscibus Marinis' [1554], in Jacques-Yves Cousteau and Yves Paccalet, *Requins: Innocents sauvages*, Paris: Robert Laffont, 1997; see also: https://artsand culture.google.com/exhibit/dgLyc3wJ1r7cIg

22 R.H. Johnson, *Sharks of Polynesia*, Singapore: Les Éditions du Pacifique, 1978.

23 *Māori Myths, Legends and Contemporary Stories*, New Zealand Ministry of Education: http://eng.mataurangamaori.tki.org.nz/ Support-materials/Te-Reo-Maori/Maori-Myths-Legends-and-Contemporary-Stories/Kawariki-and-the-shark-man

24 John D. Stevens, *Sharks*, Sydney: Murdoch Books, 1987.

25 Marcel de Serres, *Des Causes des migrations des divers animaux et particulièrement des oiseaux et des poissons*, Paris: Lagny Frères, 1845.

26 Victor Hugo, *Toilers of the Sea*, trans. W. Moy Thomas, Project Gutenberg eBook, 2010, p. 291.

27 Verne, *Twenty Thousand Leagues under the Sea*.

28 *Le Petit Parisien illustré*, *Le Petit Journal illustré*.

29 Richard G. Fernicola, *Twelve Days of Terror*, Guilford, CT: Lyons Press, 2001.

30 Hergé, *Red Rackham's Treasure*, London: Methuen, 1959.

31 Ernest Hemingway, *The Old Man and the Sea*, New York: Vintage, pp. 100–1.

32 Cousteau and Cousteau, *The Shark*, pp. 17, 20.

33 Note the exaggeration in the commentary: no white shark has ever measured ten metres; they are never over seven metres.

34 B.J., Cinéma, 'Bleue est la mer, blanche est la mort', *Le Monde*, 16 February 1972. See www.lemonde.fr/archives/article/1972/02/16/cinema-bleue-est-la-mer-blanche-est-la-mort_3032814_18 19218.html

35 François Sarano, 'Grand Blanc', *Plongeurs international*, No. 33, 2000.

36 Scientists who believe that behaviour is solely conditioned by reflexive mechanisms in response to environmental stimuli.

37 Culum Brown, Kevin Laland and Jens Krause, *Fish Cognition and Behavior*, 2nd edn, London: Blackwell, 2011.

38 Andrew Stanton and Lee Unkrich, dir., *Finding Nemo*, Walt Disney Productions, 2003.

39 François Sarano, *Rencontres sauvages: Réflexion sur 40 ans d'observations sous-marines*, Challes-les-Eaux: Éditions Gap, 2011, p. 115.

40 Coralie Schaub, *François Sarano: Réconcilier les hommes avec la vie sauvage*, Arles: Actes Sud, 2020, pp. 122–8.

41 https://youtu.be/XqZsoesa55w

Chapter 2: Shark? What Shark?

1 Heekyung Jung, et al., 'The Ancient Origins of Neural Substrates for Land Walking', *Cell*, 172/4, 2018, pp. 667–82.

2 Said of a part of the body that is not in the right place, and by extension, anything that is not in its proper milieu.

3 Romain Vullo, et al., 'Manta-Like Planktivorous Sharks in Late Cretaceous Oceans', *Science*, 371/6535, 2021, pp. 1253–6.

4 Yuichiro Hara, et al., 'Shark Genomes Provide Insights into Elasmobranch Evolution and the Origin of Vertebrates', *Nature, Ecology and Evolution*, 2/11, 2018, pp. 1761–71.

5 Pliny, *Natural History*.

6 Byrappa Venkatesh, et al., 'Elephant Shark Genome Provides

Unique Insights into Gnathostome Evolution, *Nature*, 505, 2014, pp. 174–9.

7 Romain Gary, *The Roots of Heaven*, trans. Jonathan Griffin, New York: Pocket Books, 1958 [1956], p. 375.

8 See the Visible Paleo-Earth, https://phl.upr.edu/projects/visible-paleo-earth

9 Jaw-less.

10 Min Zhu, et al., 'A Silurian Placoderm with Osteichthyan-like Marginal Jaw Bones', *Nature*, 502/7470, 2013, pp. 188–93.

11 Vertebrates with a jaw.

12 Venkatesh et al., 'Elephant Shark Genome'.

13 Gilles Cuny and Alain Bénéteau, *Requins: De la préhistoire à nos Jours*, Paris: Belin, 2013, p. 16.

14 Ibid., p. 30.

15 Zhu et al., 'A Silurian Placoderm'.

16 Robert Carr, 'Paleoecology of Dunkleosteus Terrelli (placodermi: arthrodira)', *Kirtlandia: The Cleveland Museum of Natural History*, 57, 2010, pp. 36–45.

17 Cuny and Bénéteau, *Requins*.

18 Ibid., p. 48.

19 Ibid., p. 104.

20 Jason B. Ramsay, et al., 'Eating with a Saw for a Jaw: Functional Morphology of the Jaws and Tooth-Whorl in *Helicoprion davisii*', *Journal of Morphology*, 276/1, 2015, pp. 47–64.

21 Homer, *The Odyssey*, Book XII, trans. Emily Wilson, New York: Norton, 2018, p. 304.

22 One knot is equivalent to 1.85 kilometres per hour; four knots to 7.4 kilometres per hour, the speed of the world record fifty-metre swimmer.

23 Cuny and Bénéteau, *Requins*, pp. 23 and 146.

24 François Sarano and Stéphane Granzotto, *Méditerranée, le royaume perdu des requins*, doc., Mona Lisa Production-France 2, 52 min, 2013.

25 Leighton R. Taylor, Leonard J.V. Compagno and Paul J. Struhsaker, 'Megamouth: A New Species, Genus, and Family of

Lamnoid Shark (*Megachasma pelagios*, Family *Megachasmidae*) from the Hawaiian Islands', *Proceedings of the California Academy of Sciences*, 43/8, 1983, pp. 87–110.

26 J.P. Bourseau, N. Améziane-Cominardi and M. Roux, 'Un crinoïde pédonculé nouveau (Échinodermes), représentant actuel de la famille jurassique des Hemicrinidae: *Gymnocrinus richeri*', nov. Sp. des fonds bathyaux de Nouvelle-Calédonie (S.-O. Pacifique), *Comptes rendus de l'Académie des sciences*, 305, 1987, pp. 595–9.

27 Bernard Seret, 'Découverte d'une faune à *Procarcharodon Megalodon* (Agassiz, 1835) en Nouvelle-Calédonie (*Pisces, Chondrichthyes, Lamnidae*)', *Cybium*, 11/4, 1987, pp. 389–94.

28 Jack A. Cooper, et al., 'Body Dimensions of the Extinct Giant Shark Otodus megalodon: A 2D Reconstruction', *Scientific Reports*, 10, 2020, 14596.

29 S. J. Godfrey, J. R. Nance and N. L. Riker, '*Otodus*-Bitten Sperm Whale Tooth from the Neogene of the Coastal Eastern United States', *Acta Palaeontologica Polonica*, 66/3, 2021, pp. 599–603.

30 See chapter 7, 'The Ocean is Their Garden', pp. 134–53.

31 A.N. Neumann, et al., 'The Extinction of Iconic Megatoothed Shark Otodus Megalodon: Preliminary Evidence from "Clumped" Isotope Thermometry', *American Geophysical Union*, Fall meeting, 2018.

32 Robert Flaherty, dir., *Man of Aran*, 1934, 76 min.

33 Deep water rises to the surface under the influence of the wind, which pushes the surface water out to sea. This cold, nutrient-rich deep water allows the development of planktonic algae and, consequently, the explosion of life.

34 'Fishermen in Peru catch GIANT 26ft manta ray', https://www.youtube.com/watch?v=pNbjPCYFxcI

35 Nicholas K. Dulvy, et al., 'Overfishing Drives Over One-Third of All Sharks and Rays toward a Global Extinction Crisis', *Current Biology*, 31/21, 2021, pp. 4773–87.e8.

36 David A. Ebert, Marc Dando and Sarah Fowler, *Sharks of the*

World: A Complete Guide, Princeton, NJ: Princeton University Press, 2021.

37 Sarah T.F.L. Viana and Marcelo R. de Carvalho, '*Squalus shiraii* sp. nov. (Squaliformes, Squalidae), a New Species of Dogfish Shark from Japan with Regional Nominal Species Revisited', *Zoosystematics and Evolution*, 96/2, 2020, pp. 275–311; Fahmi, Ian R. Tibbetts, Michael B. Bennett and Christine L. Dudgeon, 'Delimiting Cryptic Species within the Brown-Banded Bamboo Shark, *Chiloscyllium punctatum*, in the Indo-Australian Region with Mitochondrial DNA and Genome-Wide SNP Approaches', *BMC Ecology and Evolution*, 21, 2021, p. 121.

Chapter 3: Giving Life

1 A large marine gastropod which can weigh up to 1.5 kilos.

2 Shark Research Institute, www.sharks.org/lemon-shark-negaprion-brevirostris

3 Kevin A. Feldheim, et al., 'Two Decades of Genetic Profiling Yields First Evidence of Natal Philopatry and Long-Term Fidelity to Parturition Sites in Sharks', *Molecular Ecology*, 23/1, 2013, pp. 110–17.

4 S.P. Oliver and A.E. Bicskos, 'A Pelagic Thresher Shark (*Alopias pelagicus*) Gives Birth at a Cleaning Station in the Philippines', *Coral Reefs*, 34/1, 2014, p. 17.

5 Erich K. Ritter and Raid W. Amin, 'Mating Scars among Sharks: Evidence of Coercive Mating?', *Acta Ethologica*, 22/6, 2019, pp. 9–16.

6 Cartilaginous gutters which extend each pelvic fin and through which the semen flows.

7 P. Salinas-de-León, M. Hoyos and F. Pochet, 'First Observation on the Mating Behavior of the Endangered Scalloped Hammerhead Shark *Sphyrna lewini* in the Tropical Eastern Pacific', *Environmental Biology of Fishes*, 100, 2017, pp. 1603–8.

8 Helen Colbachini, et al., 'Body Movement as an Indicator of Proceptive Behavior in Nurse Sharks (*Ginglymostoma cirratum*)', *Environmental Biology of Fishes*, 103, 2020, pp. 1257–63.

9 Cuny and Bénéteau, *Requins*, pp. 30–1.

10 John A. Long, et al., 'Copulation in Antiarch Placoderms and the Origin of Gnathostome Internal Fertilization', *Nature*, 517, 2015, pp. 196–9.

11 Zhu et al., 'A Silurian Placoderm'.

12 Amanda M. Barker, et al., 'High Rates of Genetic Polyandry in the Blacknose Shark, *Carcharhinus acronotus*', *Copeia*, 107/3, 2019, pp. 502–8.

13 Felipe Lamarca, et al., 'Is Multiple Paternity in Elasmobranchs a Plesiomorphic Characteristic?', *Environmental Biology of Fishes*, 12, 2020, pp. 1463–70.

14 S.N. Maduna, et al., 'Evidence for Sperm Storage in Common Smoothhound Shark, *Mustelus mustelus*, and Paternity Assessment in a Single Litter from South Africa', *Journal of Fish Biology*, 92/4, 2018, p. 1183–91.

15 Jennifer V. Schmidt, et al., 'Paternity Analysis in a Litter of Whale Shark Embryos', *Endangered Species Research*, 12, 2010, pp. 117–24.

16 Ecological amnesia affects the person who considers the state of a natural environment that he or she is observing for the first time as the 'original situation'. Thus, from generation to generation, there is a shift in the 'original' reference, which means that the evolution felt by an individual during his or her lifetime does not correspond to the whole of the anthropic upheavals which affect the environment over several hundred years.

17 Emmanuelle Pouydebat and Julie Terrazzoni, *Sexus animalus: Tous les goûts sont dans la nature*, Paris: Arthaud, 2020.

18 Development of an individual from an unfertilized egg. Observed in insects, worms, amphibians, certain species of monitor and sharks: the zebra shark, or the bonnethead shark.

19 Christine L. Dudgeon, et al., 'Switch from Sexual to Parthenogenetic Reproduction in a Zebra Shark', *Scientific Reports*, 7, 2017, 40537.

20 Name given to the geological era that is undergoing deep alterations due to industrial activity. Man is the main actor in global

change. Theorized by the 1995 Nobel Prize winner in Chemistry, Paul Josef Crutzen.

21 Lisa J. Natanson and Gregory B. Skomal, 'Age and Growth of the White Shark, *Carcharodon carcharias*, in the Western North Atlantic Ocean', *Marine and Freshwater Research*, 66/5, 2015, pp. 387–98.

22 Julius Nielsen, et al., 'Eye Lens Radiocarbon Reveals Centuries of Longevity in the Greenland Shark (*Somniosus microcephalus*)', *Science*, 353/6300, 2016, pp. 702–4.

23 Christopher G. Mull, Kara E. Yopak and Nicholas K. Dulvy, 'Does More Maternal Investment Mean a Larger Brain? Evolutionary Relationships between Reproductive Mode and Brain Size in Chondrichthyans', *Marine and Freshwater Research*, 62/6, 2011, pp. 567–75.

24 Schaub, *François Sarano*, pp. 122–8.

Chapter 4: Inside the Shark's Head

1 On board *Calypso*, we used waterproof underwater cameras loaded with 16-millimetre film. A motor was used to wind the film onto the take-up reel. Video and digital cameras did not yet exist.

2 Bernard Quéguiner and Peggy Rimmelin, 'Biogenic and Lithogenic Particulate Silica Collected during the Biosope (2004) Cruise', 2018, SEANO, https://doi.org/10.17882/55722

3 Patrick Süskind, *Perfume: The Story of a Murderer*, New York: A.A. Knopf, 1986.

4 Hemingway, *The Old Man and the Sea*, p. 99.

5 Jayne M. Gardiner, 'Multisensory Integration in Shark Feeding Behavior', PhD, University of South Florida, 2012, chap. 1, p. 1, https://digitalcommons.usf.edu/etd/4046/

6 The stream of particles loaded with smells enters the inhalation channels of each of the two nostrils, passes over a bed of sensory cells – sensitive to minute concentrations – and exits through the exhalation channel. Unlike sound waves, which propagate fairly evenly through water, odour molecules are diluted and dispersed

chaotically in the flow. Large puffs of highly concentrated molecules and small ones, a hundred times less concentrated, are distributed randomly.

7 Gardiner, 'Multisensory Integration', p. 68.

8 Eric M. Stroud, et al., 'Chemical Shark Repellent: Myth or Fact? The Effect of a Shark Necromone on Shark Feeding Behavior', *Ocean and Coastal Management*, 97, 2014, pp. 50–7.

9 Molecules of oleic acid, released shortly before death.

10 Kara Yopak, T. Lisney and S. Collin, 'Not All Sharks Are "Swimming Noses": Variation in Olfactory Bulb Size in Cartilaginous Fishes', *Brain Structure & Function*, 220/2, 2015, pp. 1127–43.

11 Richard H. Johnson and Donald R. Nelson, 'Copulation and Possible Olfaction-Mediated Pair Formation in Two Species of Carcharhinid Sharks', *Copeia*, 1978, pp. 539–42; Johnson, *Sharks of Polynesia*, p. 39.

12 Kurt Kotrschal, 'Ecomorphology of Solitary Chemosensory Cell Systems in Fish: A Review', *Environmental Biology of Fishes*, 44, 1995, pp. 143–55; Meredith B. Peach, 'New Microvillous Cells with Possible Sensory Function on the Skin of Sharks', *Marine and Freshwater Behavior and Physiology*, 38/4, 2005, pp. 275–9.

13 Fish also benefit from the same kind of receptive system.

14 Gardiner, 'Multisensory Integration', chap. 1, p. 14.

15 Jayne M. Gardiner and J. Atema, 'Sharks Need the Lateral Line to Locate Odor Sources: Rheotaxis and Eddy Chemotaxis', *Journal of Experimental Biology*, 210/11, 2007, pp. 1925–34.

16 Donald R. Nelson, and S.H. Gruber, 'Sharks: Attraction by Low-Frequency Sounds', *Science*, 142/3594, 1963, pp. 975–7.

17 Gardiner, 'Multisensory Integration', chap. 1, p. 17.

18 Ibid., p. 19.

19 Ibid.

20 See the classification of the orders of shark species on p. 56.

21 Stefano Lorenzini, Italian physician and ichthyologist of the late seventeenth century.

22 Kanika Sharma, et al., 'The Chemosensory Receptor Repertoire

of a True Shark Is Dominated by a Single Olfactory Receptor Family', *Genome Biology and Evolution*, 11/2, 2019, pp. 398–405.

23 Gardiner, 'Multisensory Integration', Abstract, p. viii.

24 Phenomenal consciousness is the felt, 'lived' consciousness, associating a qualitative whole (linked to the personal memorized experience) to 'cognitive' consciousness, or perception, of physical phenomena.

25 Thomas Nagel, 'What Is It Like to Be a Bat?', *The Philosophical Review*, 83/4, 1974, pp. 435–50.

Chapter 5: On the Road to Personality

1 Also known as Carlota.

2 See the QR code showing the morphological characteristics that identify each bull shark.

3 Like us humans, every organism living today is the result of 3.8 billion years of evolution. Therefore, we are all equally evolved, although we are all different and more or less complex. See Schaub, *François Sarano*, p. 128.

4 Eric Clua and Dennis Reid, 'Features and Motivation of a Fatal Attack by a Juvenile White Shark, *Carcharodon carcharias*, on a Young Male Surfer in New Caledonia (South Pacific)', *Journal of Forensic and Legal Medicine*, 20/5, 2013, pp. 551–4.

5 François Sarano, 'La raie amoureuse', *Calypsolog*, 94, 1990.

6 François Sarano, *Le Retour de Moby Dick*, Arles: Actes Sud, Mondes sauvages series, 2017, pp. 100–14.

7 Gordon M. Burghardt, 'Creativity, Play, and the Pace of Evolution', in Allison B. Kaufman and James C. Kaufman, eds., *Animal Creativity and Innovation*, Amsterdam: Elsevier, 2015, pp. 129–61.

8 Vera Schluessel, 'Who Would Have Thought That Jaws Also Have Brains? Cognitive Functions in Elasmobranchs', *Animal Cognition*, 18/1, 2015, pp. 19–37.

9 The part of the brain that joins the cerebral hemispheres.

10 Csilla Ari, 'Encephalization and Brain Organization of Mobulid Rays (*Myliobatiformes, Elasmobranchii*) with Ecological

Perspectives', *The Open Anatomy Journal*, 3/1, 2011, pp. 1–13.

11 Sarah Keartes, 'Massive Oceanic Manta Ray Accidentally Caught in Peru', *Earth Touch*, 27 April 2015.

12 Appendages that extend from the gill arch (but are distinct from the gill filaments, which are used for gas exchange) and are involved in the filtration of tiny prey suspended in the water.

13 Robert Perryman, et al., 'Social Preferences and Network Structure in a Population of Reef Manta Rays', *Behavioral Ecology and Sociobiology*, 114, 2019, pp. 1–18.

14 Cousteau and Cousteau, *The Shark*, p. 32.

15 E.E. Byrnes and C. Brown, 'Individual Personality Differences in Port Jackson Sharks, *Heterodontus portusjacksoni*', *Journal of Fish Biology*, 89/2, 2016, pp. 1142–57.

16 J.S. Finger, et al., 'Rate of Movement of Juvenile Lemon Sharks in a Novel Open Field: Are We Measuring Activity for Reaction to Novelty?', *Animal Behavior*, 116, 2016, pp. 75–82.

17 Johann Mourier, Julie Vercelloni and Serge Planes, 'Evidence of Social Communities in a Spatially Structured Network of a Free-Ranging Shark Species', *Animal Behavior*, 83/2, 2012, pp. 389–401.

18 Sarano, *Le Retour de Moby Dick*.

19 Johann Mourier, Culum Brown and Serge Planes, 'Learning and Robustness to Catch-and-Release Fishing in a Shark Social Network', *Biology Letters*, 13/3, 2017, 20160824; Johann Mourier and Serge Planes, 'Kinship Does Not Predict the Structure of a Shark Social Network', *Behavioral Ecology*, 32/2, 2021, pp. 211–22; Mourier et al., 'Evidence of Social Communities'.

20 Vera Schluessel and Theodora Fuss, 'Something Worth Remembering: Visual Discrimination in Sharks', *Animal Cognition*, 18/2, 2015, pp. 463–71.

21 Mourier et al., 'Learning and Robustness'.

22 Lisa Kerr, et al., 'Investigations of Δ14C, δ13C, and δ15N in Vertebrae of White Shark (*Carcharodon carcharias*) from the Eastern North Pacific Ocean', *Environmental Biology of Fishes*, 77/3, 2006, pp. 337–53; Eric Clua and John Linnell, 'Individual

Shark Profiling: An Innovative and Environmentally Responsible Approach for Selectively Managing Human Fatalities', *Conservation Letters*, 12/9, 2018, pp. 1–7.

23 James A. Estrada, et al., 'Use of Isotopic Analysis of Vertebrae in Reconstructing Ontogenetic Feeding Ecology in White Sharks', *Ecology*, 87/4, 2006, pp. 829–34.

24 Clua and Reid, 'Features and Motivation of a Fatal Attack'.

25 David M.P. Jacoby, et al., 'Shark Personalities? Repeatability of Social Network Traits in a Widely Distributed Predatory Fish', *Behavioral Ecology and Sociobiology*, 68/12, 2014, pp. 1995–2003; J.S. Finger, F. Dhellemmes and T.L. Guttridge, 'Personality in Elasmobranchs with a Focus on Sharks: Early Evidence, Challenges, and Future Directions', in Jennifer Vonk, Alexander Weiss and Stan A. Kuczaj, eds., *Personality in Nonhuman Animals*, Cham: Springer International Publishing, 2017, pp. 129–52.

26 Kátya Gisela Abrantes and Adam Barnett, 'Intrapopulation Variations in Diet and Habitat Use in a Marine Apex Predator, the Broadnose Sevengill Shark (*Notorynchus cepedianus*)', *Marine Ecology Progress Series*, 442, 2011, pp. 133–48.

27 Philip Matich, et al., 'Inter-individual Differences in Ontogenetic Trophic Shifts among Three Marine Predators', *Oecologia*, 189, 2019, pp. 621–36.

28 A phenomenon whereby two groups of individuals of the same species (i.e., who have fertile offspring when they reproduce together) gradually become isolated until they can no longer reproduce together. This isolation may be geographical or cultural, or linked to anatomical or physiological changes.

29 Burghardt, 'Creativity, Play, and the Pace of Evolution'.

30 J.S. Finger, et al., 'Are Some Sharks More Social than Others? Short and Long-Term Consistencies in Social Behavior of Juvenile Lemon Sharks', *Behavioral Ecology and Sociobiology*, 72, 2018, 17.

31 Sarano, *Le Retour de Moby Dick*, pp. 117–20.

32 The mirror test is a test to highlight an individual's self-awareness.

It consists of placing the individual in front of a mirror, after having painted, without their knowledge, a visible mark on their body. If the individual reacts and tries to remove this added mark, it means that he or she recognizes his or her image in the mirror and is therefore self-aware.

33 Csilla Ari and Dominic P. D'Agostino, 'Contingency Checking and Self-Directed Behaviors in Giant Manta Rays: Do Elasmobranchs Have Self-Awareness?', *Journal of Ethology*, 34/2, 2016, pp. 167–74.

34 Pierre Le Neindre, 'Conscience des animaux: Quels consensus scientifique?', *Sesame*, 6, interview with S. Berthier, 2019.

35 Pain is the physical perception of a bodily alteration and suffering is psychological. There can be suffering without physical alteration, i.e. without pain.

36 J. David Smith, et al., 'The Uncertain Response in the Bottlenosed Dolphin (*Tursiops truncatus*)', *Journal of Experimental Psychology General*, 124/4, 1995, pp. 391–408.

37 Tristan Guttridge, et al., 'Social Learning in Juvenile Lemon Sharks, *Negaprion brevirostris*', *Animal Cognition*, 16/1, 2013, pp. 55–64.

38 Catarina Vila Pouca, et al., 'Social Learning in Solitary Juvenile Sharks', *Animal Behavior*, 159, 2019, pp. 21–7.

Chapter 6: The Shark, Where it Belongs

1 Animals that feed on plant or animal plankton. Most often small fish, such as sardines or anchovies, or sometimes giants such as the basking shark or manta ray.

2 Anthropogenic factors such as fishing and pollution are not taken into account here.

3 El Niño: cyclical phenomenon corresponding to the sudden return, towards the east of the Pacific Ocean (South America), of surface waters that have been pushed westwards by the trade winds (towards Australia). This mass of warm water then covers the cold Humboldt Current, which normally carries nutrients up from the Antarctic along the coasts of Chile and Peru. As

a result, the nutrient salts remain deep in the ocean, far from the sunlit area where the algae live, so they cannot use them for photosynthesis. This phenomenon affects the entire global atmospheric and oceanic circulation.

4 Daniel M. Ware and Richard E. Thomson, 'Bottom-Up Ecosystem Trophic Dynamics Determine Fish Production in the Northeast Pacific', *Science*, 308/5726, 2005, pp. 1280–4; Richard T. Barber and Francisco P. Chavez, 'Biological Consequences of El Niño', *Science*, 222/4629, 1983, pp. 1203–10; Rachel. A. Skubel, et al., 'Patterns of Long-Term Climate Variability and Predation Rates by a Marine Apex Predator, the White Shark *Carcharodon carcharias*', *Marine Ecology Progress Series*, 587, 2018, pp. 129–39.

5 Jayson M. Semmens, et al., 'Feeding Requirements of White Sharks May Be Higher than Originally Thought', *Scientific Reports*, 3, 2013, 1471.

6 www.youtube.com/watch?v=b-m68A-7bQs&feature=youtu.be; www.youtube.com/watch?v=PxADeN_cyv0&feature=youtu.be

7 Francis G. Carey, et al., 'Temperature and Activities of a White Shark, *Carcharodon carcharias*', *Copeia*, 1982, pp. 254–60.

8 Enric Cortés and Samuel H. Gruber, 'Diet, Feeding Habits and Estimates of Daily Ration of Young Lemon Sharks, *Negaprion brevirostris* (Poey)', *Copeia*, 1990, pp. 204–18.

9 Atlantic Dawn Trawler: https://britishseafishin.co.uk/atlantic-dawn-the-ship-from-hell

10 Barry Bruce, et al., 'A National Assessment of the Status of White Sharks', *National Environmental Science Programme*, Marine Biodiversity Hub, CSIRO, Hobart, Tasmania, 2018.

11 C.E. Stillwell and N.E. Kohler, 'Food, Feeding Habits and Estimates of Daily Ration of the Shortfin Mako (*Isurus oxyrinchus*) in the Northwest Atlantic', *Journal canadien des sciences halieutiques et aquatiques*, 39/3, 1982, pp. 407–14.

12 Anabela Maia, et al., 'Food Habits of the Shortfin Mako, *Isurus oxyrinchus*, off the Southwest Coast of Portugal', *Environmental Biology of Fishes*, 77/2, 2006, pp. 157–67.

13 Renny Harlin, dir., *Deep Blue Sea*, Warner Bros. and Village Roadshow Pictures, 1999.

14 Matthieu Coutant, La Rochelle aquarium, personal communication, 2021.

15 Stéphanie Orengo, Institut océanographique de Monaco, personal communication, 2021.

16 Chelsea M. Brown, et al., 'Short-Term Changes in Reef Fish Community Metrics Correlate with Variability in Large Sharks Occurrence', *Food Webs*, 24/7, 2020, e00147; Emily K. Lester, et al., 'Relative Influence of Predators, Competitors and Seascape Heterogeneity on Behavior and Abundance of Coral Reef Mesopredators', *Oikos: Advancing Ecology*, 130/12, 2021, pp. 2239–49.

17 A caudal fin with very unequal lobes is called 'heterocercal', as opposed to 'homocercal' when the lobes are equal.

18 This means more predators than potential prey.

19 Laurent Ballesta, *700 requins dans la nuit*, Mauguio: Andromède Océanologie, 2017.

20 Johann Mourier, et al., 'Extreme Inverted Trophic Pyramid of Reef Sharks Supported by Spawning Groupers', *Current Biology*, 26/15, 2016, 2011–16.

21 Average weight of a reproductive grouper. Laurent Debas, personal communication, and doctoral thesis: 'Étude biométrique, histologique et endocrinologique de la sexualité du mérou *Epinephelus microdon* dans le milieu naturel et en élevage: Caractérisation de l'hermaphrodisme protérogyne, description du phénomène d'inversion et mise en évidence du phénomène de reversion', www.theses.fr/1989AIX22077

22 Andrés F. Navia, et al., 'How Many Trophic Roles Can Elasmobranchs Play in a Marine Tropical Network?', *Marine and Freshwater Research*, 68/7, 2016, pp. 1–12.

23 Tristan L. Guttridge, et al., 'Deep Danger: Intraspecific Predation Risk Influences Habitat Use and Aggregation Formation of Juvenile Lemon Shark', *Marine Ecology Progress Series*, 445, 2012, pp. 279–91; Rafael Tavares, 'Survival Estimates of Juvenile

Lemon Sharks Based on Tag-Recapture Data at Los Roques Archipelago, Southern Caribbean', *Caribbean Journal of Science*, 50/1, 2020, pp. 171–7.

24 Jayson M. Semmens, et al., 'Feeding Requirements of White Sharks May Be Higher than Originally Thought', *Scientific Reports*, 3, 2013, 1471.

25 Alexandra C. Aines, et al., 'Feeding Habits of the Tiger Shark, *Galeocerdo cuvier*, in the Northwest Atlantic Ocean and Gulf of Mexico', *Environmental Biology of Fishes*, 101, 2018, pp. 403–15.

26 Jenna L. Hounslow, et al., 'Animal-Borne Video from a Sea Turtle Reveals Novel Anti-Predator Behaviors', *Ecology*, 102/4, 2021, e03251.

27 Michael R. Heithaus, et al., 'State-Dependent Risk-Taking by Green Sea Turtles Mediates Top-Down Effects of Tiger Shark Intimidation in a Marine Ecosystem', *Journal of Animal Ecology*, 76, 2007, pp. 837–44.

28 WWF: www.worldwildlife.org/stories/five-way-sharks-and-rays-help-the-world

29 Sea Shepherd: www.seashepherd.nc/protection-du-requin-amelen

30 A direct downward chain reaction whereby the removal of one species from a given trophic level leads to a population explosion of species in the trophic level directly below.

31 William J. Ripple and Robert L. Beschta, 'Large Predators Limit Herbivore Densities in Northern Forest Ecosystems', *European Journal of Wildlife Research*, 58, 2012, pp. 733–42; Ramana Callan, et al., 'Recolonizing Wolves Trigger a Trophic Cascade in Wisconsin (USA)', *Journal of Ecology*, 101/4, 2013, pp. 837–45.

32 David Shiffman, 'Sharks Create Oxygen: A Scientific Perspective', *Southern Fried Science*, 10 February 2012.

33 The food pyramid is a diagram which stacks the trophic levels that structure the energy relationships between species. This pyramid explains that it takes ten kilos of plants for a herbivore to grow by one kilo; ten kilos of herbivores for a first-level carni-

vore to grow by one kilo; and so on up to the carnivorous apex predators of the third or fourth level.

34 Ransom A. Myers, et al., 'Cascading Effects of the Loss of Apex Predatory Sharks from a Coastal Ocean', *Science*, 315/5820, 2007, pp. 1846–50.

35 Ware and Thomson, 'Bottom-Up Ecosystem Trophic Dynamics'.

36 Sharks exert their predation pressure on fish that are large enough to be 'visible' and 'edible', i.e. large immature and adult fish, which represent only 1/100,000 of the number of eggs that are released into the marine environment, i.e. an extremely narrow fraction of the initial population. They do not exert any predation pressure on the eggs, larvae and juveniles, which are 100,000 times more numerous. On the other hand, it is on this enormous potential source of eggs and larvae that the predominant influence of hydroclimatic physico-chemical variations and that of small plankton eaters (copepods, jellyfish and other chaetognaths) is exerted. A very small variation in hydroclimatic constraints can therefore considerably vary the success of reproduction by a factor of around ten. From that point on, all bets are off, regardless of how much the large predators take, because it cannot vary by this order of magnitude so quickly. This is why, for example, in the Bay of Biscay, there are considerable annual variations in anchovy populations, of the order of one to ten billion individuals, independently of the catches of their predators, mackerel, tuna and sharks. Global warming, which disrupts the temperature and consequently the marine currents, has an infinitely greater influence on the oceanic populations than that exerted by predators playing on the sidelines.

37 Navia et al., 'How Many Trophic Roles Can Elasmobranchs Play'; Luciana C. Ferreira, et al., 'The Trophic Role of a Large Marine Predator, the Tiger Shark *Galeocerdo cuvier*', *Scientific Reports*, 7/1, 2017, pp. 1–14; Daniel J. Madigan, et al., 'Stable Isotope Analysis Challenges Wasp-Waist Food Web Assumptions in an Upwelling Pelagic Ecosystem', *Scientific Reports*, 2, 2012, 654.

38 Sora L. Kim, et al., 'Ontogenetic and Among-Individual Variation

in Foraging Strategies of Northeast Pacific White Sharks Based on Stable Isotope Analysis', *PLoS One*, 7/9, 2012, e45068; Matich, et al., 'Inter-individual Differences in Ontogenetic Trophic Shifts'.

39 Nigel E. Hussey, et al., 'Expanded Trophic Complexity among Large Sharks', *Food Webs*, 4, 2015, pp. 1–7.

40 Jeremy J. Kiszka, et al., 'Plasticity of Trophic Interactions among Sharks from the Oceanic South-Western Indian Ocean Revealed by Stable Isotope and Mercury Analyses', *Deep Sea Research Part I*, 96, 2015, pp. 49–58.

41 Paul L. Rogers, et al., 'A Quantitative Comparison of the Diets of Sympatric Pelagic Sharks in Gulf and Shelf Ecosystems off Southern Australia', *ICES Journal of Marine Science*, 69/8, 2012, pp. 1382–93.

42 Ashley J. Frisch, et al., 'Reassessing the Trophic Role of Reef Sharks as Apex Predators on Coral Reefs', *Coral Reefs*, 35, 2016, pp. 459–72.

43 Philip Matich, et al., 'Species Co-occurrence Affects the Trophic Interactions of Two Juvenile Reef Shark Species in Tropical Lagoon Nurseries in Moorea (French Polynesia)', *Marine Environmental Research*, 127, 2017, pp. 84–91; Schaub, *François Sarano*, p. 127.

44 George Roff, et al., 'The Ecological Role of Sharks on Coral Reefs', *Trends in Ecology and Evolution*, 31/5, 2016, pp. 395–407; Brown, et al., 'Short-Term Changes in Reef Fish Community Metrics'.

45 Dr Johann Mourier, personal communication, 2021.

46 Schaub, *François Sarano*, pp. 120–41.

47 Pierre Labourgade, et al., 'Heterospecific Foraging Associations between Reef-Associated Sharks: First Evidence of Klepto-parasitism in Sharks', *Ecology*, 101/4, 2020, e01755.

48 The annual migration of sardines along the south-east coast of South Africa from Cape Agulhas to Durban in early winter is called the Sardine Run.

49 Christophe Lett, et al., 'Effects of Successive Predator Attacks on Prey Aggregations', *Theoretical Ecology*, 7/3, 2014, pp. 239–52; Andréa Thiebault, et al., 'How to Capture Fish in a School? Effect

of Successive Predator Attacks on Seabird Feeding Success', *Journal of Animal Ecology*, 85/1, 2015, pp. 157–67.

50 Reference to the last page of Paul Valéry's book *L'Idée fixe*, New York: Pantheon Books, 1965 [1933].

51 Ballesta, *700 requins dans la nuit*; Johann Mourier, et al., 'Visitation Patterns of Camouflage Groupers *Epinephelus polyphekadion* at a Spawning Aggregation in Fakarava Inferred by Acoustic Telemetry', *Coral Reefs*, 38/5, 2019, pp. 909–16.

52 Schaub, *François Sarano*, pp. 132–5.

53 Andrew Chin, Johann Mourier and Jodie L. Rummer, 'Blacktip Reef Sharks (*Carcharhinus melanopterus*) Show High Capacity for Wound Healing and Recovery Following Injury', *Conservation Physiology*, 3/1, 2015, cov062; Claudia Pogoreutz, et al., 'Similar Bacterial Communities on Healthy and Injured Skin of Black Tip Reef Sharks', *Animal Microbiome*, 1, 2019, 9.

54 Nicholas J. Marra, et al., 'White Shark Genome Reveals Ancient Elasmobranch Adaptations Associated with Wound Healing and the Maintenance of Genome Stability', *PNAS*, 116/10, 2019, pp. 4446–55.

55 William Lane and Linda Comac, *Sharks Still Don't Get Cancer: The Continuing Story of Shark Cartilage Therapy*, New York: Avery, 1996, p. 246.

56 Joel B. Finkelstein, 'Sharks Do Get Cancer: Few Surprises in Cartilage Research', *Journal of the National Cancer Institute*, 97/21, 2005, pp. 1562–3; Constance Holden, 'Sharks DO Get Cancer', *Science*, 288/5464, 2000, p. 259; Gary K. Ostrander, et al., 'Shark Cartilage, Cancer and the Growing Threat of Pseudoscience', *Cancer Research*, 64/23, 2004, pp. 8485–91.

57 Danny Morick, et al., 'Fatal Infection in a Wild Sandbar Shark (*Carcharhinus plumbeus*), Caused by *Streptococcus agalactiae*, Type Ia-ST7', *Animals*, 10/2, 2020, p. 284.

58 Aaron C. Henderson, K. Flannery and Jennifer Dunne, 'Parasites of the Blue Shark (*Prionace glauca L.*), in the North-East Atlantic Ocean', *Journal of Natural History*, 36/16, 2002, pp. 1995–2004.

59 Copepods, which number dozens of species, account for more

than 60 per cent of small planktonic crustaceans; 15 per cent of the species are parasites.

60 C.G. Hewitt, 'Eight Species of Parasitic Copepoda on a White Shark', *New Zealand Journal of Marine and Freshwater Research*, 13/1, 1979, p. 171.

61 Maria Pickering and Janine N. Caira, 'A New Hyperapolytic Species, Trilocularia Eberti sp. N. (*Cestoda: Tetraphyllidea*), from Squalus cf. Mitsukurii (Squaliformes: *Squalidae*) off South Africa with Comments on Its Development and Fecundity', *Folia Parasitologica*, 59/2, 2012, pp. 107–14.

62 Joshua K. Moyer, Jon Dodd and Duncan J. Irschick, 'Observation of a Sea Lamprey, *Petromyzon marinus*, on a Pelagic Blue Shark, *Prionace glauca*', *North-Eastern Naturalist*, 27/2, 2020, N16–N20.

63 Mauricio Hoyos, et al., 'Observation of an Attack by a Cookiecutter Shark (*Isistius brasiliensis*) on a White Shark (*Carcharodon carcharias*)', *Pacific Science*, 67, 2013, pp. 129–34.

64 Cheyanne Poe, 'Sharks and Microorganisms: A Case of Peaceful Cohabitation', *United Academics Magazine*, 17 August 2020, www.ua-magazine.com/2020/08/17/shark-microbiome-defense

65 Samantha C. Leigh, Yannis P. Papastamatiou and Donovan P. German, 'Gut Microbial Diversity and Digestive Function of an Omnivorous Shark', *Marine Biology*, 168, 2021, 55; Samantha C. Leigh, Yannis P. Papastamatiou and Donovan P. German, 'The Nutritional Physiology of Sharks', *Review in Fish Biology and Fisheries*, 27, 2017, pp. 561–85.

66 For further information on the world of indispensable microbes, see Marc-André Selosse, *Jamais seul: Ces microbes qui construisent les plantes, les animaux et les civilisations*, Arles: Actes Sud, 2017.

67 Unlike fish, whose mouth–gill movement is sufficient to induce the water current even when the fish is motionless on the bottom, most sharks have to swim actively, with their mouths open, to ensure that the water current from which the gills draw oxygen is sufficient.

68 Simon P. Oliver, et al., 'Oceanic Sharks Clean at Coastal Seamount', *PLoS One*, 6/3, 2011, e14755; Helen F. Cadwallader, J.R. Turner and Simon P. Oliver, 'Cleaner Wrasse Forage on Ectoparasitic Digeneans (Phylum Platyhelminthes) that Infect Pelagic Thresher Sharks (*Alopias pelagicus*)', *Marine Biodiversity*, 45/4, 2014, pp. 613–14.

69 Redouan Bshary and Isabelle Côté, 'New Perspectives on Marine Cleaning Mutualism', in *Fish Behaviour*, Boca Raton, FL: CRC Press, 2008, pp. 563–92.

70 Peter A. Waldie, et al., 'Long-Term Effects of the Cleaner Fish *Labroides dimidiatus* on Coral Reef Fish Communities', *PLoS One*, 6/6, 2011, e21201.

71 Bshary and Côté, 'New Perspectives on Marine Cleaning Mutualism'.

72 Perrin and Cluzaud, *Oceans*.

73 Generic term for several species of large brown algae, of the Laminariaceae family, which can measure up to several dozen metres.

74 Neil Hammerschlag, et al., 'Disappearance of White Sharks Leads to the Novel Emergence of an Allopatric Apex Predator, the Sevengill Shark', *Scientific Reports*, 9, 2019, 1908.

75 Tamlyn M. Engelbrecht, Alison A. Kock and M. Justin O'Riain, 'Running Scared: When Predators Become Prey', *Ecosphere*, 10/1, 2019, e02531.

76 'Shark Free Chips' campaign, https://sharkfreechips.com

77 The Cousteau team's second ship, powered by turbosails.

78 Donavan Smith, 'Killer whales VS Great white shark': www.youtube.com/watch?v=Pq-T0Q91I1U; orcas chasing a tiger shark: www.youtube.com/watch?v=uqimOYOQjJ8 [video no longer available]

79 Salvador J. Jorgensen, et al., 'Killer Whales Redistribute White Shark Foraging Pressure on Seals', *Scientific Reports*, 9, 2019, 6153.

80 Craig R. Smith and Amy R. Baco, 'Ecology of Whale Falls at the Deep-Sea Floor', *Oceanography and Marine Biology: An*

Annual Review, 41, 2003, pp. 311–54; Joan M. Alfaro-Lucas, et al., 'Trophic Structure and Chemosynthesis Contributions to Heterotrophic Fauna Inhabiting an Abyssal Whale Carcass', *Marine Ecology Progress Series*, 596, 2018, pp. 1–12.

81 Nicholas D. Higgs, et al., 'Fish Food in the Deep Sea: Revisiting the Role of Large Food-Falls', *PLoS One*, 9/5, 2014, e96016.

Chapter 7: The Ocean is Their Garden

1 Paris-Montsouris meteorological station: http://meteo-climat -bzh.dyndns.org/metannee-1-1748.php

2 Vesuvius buried Pompei in 79 BCE.

3 Julius Nielsen, et al., 'Eye Lens Radio-Carbon Reveals Centuries of Longevity in the Greenland Shark (*Somniosus microcephalus*)', *Science*, 353/6300, 2016, pp. 702–4.

4 French shark study group in Greenland: https://geerg.ca/fr/la-verite-sur-le-requin-du-groenland

5 Programme TOPP (Tagging of Pelagic Predators): https://gtopp .org/about-gtopp/animals/salmon-sharks.html

6 *Lamna ditropis* maintains its body temperature by means of the *rete mirabile*, an extremely well-developed blood network that recovers heat from muscle contractions.

7 Aaron B. Carlisle, et al., 'Stable Isotope Analysis of Vertebrae Reveals Ontogenetic Changes in Habitat in an Endothermic Pelagic Shark', *Proceedings of the Royal Society B: Biological Sciences*, 282/1799, 2014, 20141446.

8 These are the great white sharks of the Mediterranean that Marcel de Serres called *Squalus carcharias*.

9 De Serres, *Des causes des migrations des divers animaux*, p. 397.

10 Elder son of Captain Jacques-Yves Cousteau.

11 Jean-Michel Cousteau and Mose Richards, *Cousteau's Great White Shark*, New York: Harry N. Abrams, 1992, p. 160.

12 Russell W. Bradford, et al., 'Evidence of Diverse Movement Strategies and Habitat Use by White Sharks, *Carcharodon*

carcharias, off Southern Australia', *Marine Biology*, 167, 2020, 96.

13 Ramón Bonfil, et al., 'Transoceanic Migration, Spatial Dynamics, and Population Linkages of White Sharks', *Science*, 310/5745, 2005, pp. 100–3.

14 Michael L. Domeier, Nicole Nasby-Lucas and Chi H. Lam, 'Fine-Scale Habitat Use by White Sharks at Guadalupe Island, Mexico', in Michael L. Domeier, ed., *Global Perspectives on the Biology and Life History of the White Shark*, Boca Raton, FL: CRC Press, 2012, pp. 121–32.

15 Kevin C. Weng, et al., 'Migration and Habitat of White Sharks (*Carcharodon carcharias*) in the Eastern Pacific Ocean', *Marine Biology*, 152, 2007, pp. 877–94.

16 Gaël Le Croizier, et al., 'The Twilight Zone as a Major Foraging Habitat and Mercury Source for the Great White Shark', *Environmental Science and Technology*, 54/24, 2020, pp. 15872–82.

17 The area between 200 and 1,000 metres in depth that separates the sunlit zone from the permanent night zone.

18 Deep Blue in Oahu (Hawaii), 14 January, 2019, www.youtube.com/watch?v=KMRK0sDhmi0

19 Michael L. Domeier and Nicole Nasby-Lucas, 'Two-Year Migration of Adult Female White Sharks (*Carcharodon carcharias*) Reveals Widely Separated Nursery Areas and Conservation Concerns', *Animal Biotelemetry*, 1, 2013, 2.

20 Julia L.Y. Spaet, et al., 'Spatiotemporal Distribution Patterns of Immature Australasian White Sharks (*Carcharodon carcharias*)', *Scientific Reports*, 10, 2020, 10169.

21 Gregory B. Skomal, et al., 'Movements of the White shark *Carcharodon carcharias* in the North Atlantic Ocean', *Marine Ecology Progress Series*, 580, 2017, pp. 1–16.

22 Nukimi, an adventurous great white shark, fifty years old and two tonnes, is being followed through the north Atlantic by *Ocearch*, www.ocearch.org/tracker/detail/nukumi

23 Exhibition at the Lausanne Zoological Museum. François Sarano and Stéphane Granzotto, *Méditerranée*.

24 Chrysoula Gubili, et al., 'Antipodean White Sharks on a Mediterranean Walkabout? Historical Dispersal Leads to Genetic Discontinuity and an Endangered Anomalous Population', *Proceedings of the Royal Society B. Biological Sciences*, 278/1712, 2011, pp. 1679–86.

25 Camilo Montes, et al., 'Middle Miocene Closure of the Central American Seaway', *Science*, 348/6231, 2015, pp. 226–9.

26 Salvador J. Jorgensen, et al., 'Philopatry and Migrations of Pacific White Sharks', *Proceedings of the Royal Society B. Biological Sciences*, 277/1682, 2009, pp. 679–88.

27 Alberto Collareta, et al., 'Until Panama Do Us Part: New Finds from the Pliocene of Ecuador Provide Insights into the Origin and Palaeobiogeographic History of the Extant Requiem Sharks *Carcharhinus acronotus* and *Nasolamia velox*', *Neues Jahrbuch für Geologie und Paläontologie Abhandlungen*, 300, 2021, pp. 103–15.

28 Said of two species or populations that live in the same place without hybridizing.

29 See chapter 5 of this book.

30 François Sarano, *Rencontres sauvages: Réflexion sur 40 ans d'observation sous-marines*, Gap, 2011, pp. 212–19.

31 Gregory B. Skomal, et al., 'Transequatorial Migrations by Basking Sharks in the Western Atlantic Ocean', *Current Biology*, 19/12, 2009, pp. 1019–22.

32 For illustrations of different planetary projections, including an oceanographic projection, see Athelstan Spilhaus, http://cartonu merique.blogspot.com/2020/09/projection-spilhaus.html

33 Continental drift over 750 million years, https://sites.google.com /a/upr.edu/planetary-habitability-laboratory-upra/projects/visu al-paleo-earth/vpe-animations

34 Below 4°C, the amount of salt that water can dissolve decreases, and density is a function of the amount of dissolved salt.

35 Global circulation of ocean waters related to differences in density

– temperature (thermo) and salinity (haline), www.youtube
.com/watch?v=nhmposeF2gk

36 James M. Anderson, et al., 'Insight into Shark Magnetic Field
Perception from Empirical Observations', *Scientific Reports*, 7,
2017, 11042.

37 Bryan A. Keller, et al., 'Map-Like Use of Earth's Magnetic Field
in Sharks', *Current Biology*, 31/13, 2021, pp. 2881–6; Carl G.
Meyer, Kim N. Holland and Yannis P. Papastamatiou, 'Sharks
Can Detect Changes in the Geomagnetic Field', *Journal of the
Royal Society Interface*, 2/2, 2005, pp. 129–30.

38 See the European Space Agency (ESA): www.esa.int/Appl
ications/Observing_the_Earth/Swarm/Swarm_tracks_elusive
_ocean_magnetism

39 Carl G. Meyer and Kim N. Holland, 'The Shark Magnetic Sense',
Hawaii Institute of Marine Biology, Shark Research, www.himb
.hawaii.edu/ReefPredator/Shark%20Magnet.htm

40 www.notre-planete.info/actualites/1754-champ-magnetique-in
version-poles

41 Philip W. Livermore, C. Finlay and M. Bayliff, 'Recent North
Magnetic Pole Acceleration towards Siberia Caused by Flux
Lobe Elongation', *Nature Geoscience*, 13, 2020, pp. 387–91.

42 Ibid.

43 The opportunistic diversion of the primary function of an organ
or behaviour to another use that gradually becomes dominant.

44 David A. Mann and Phillip S. Lobel, 'Acoustic Behavior of the
Damselfish *Dascyllus albisella*: Behavioral and Geographic
Variation', *Environmental Biology of Fishes*, 51/4, 1998,
pp. 421–8.

45 A specimen whose description is used to define a new species.

46 A contact surface between two bodies of water whose difference
in temperature causes a difference in density that prevents them
from mixing. Above the thermocline, the water is warm; below it
is cold, often by several degrees. This difference in temperature is
often accompanied by a difference in salinity, which also causes

a separation between the water masses, known as the halocline. Salinity and temperature, the two main actors in the density of a water mass, lead to the formation of a pycnocline, the contact surface between two water masses of different densities.

47 Camilo Montes, et al., 'Middle Miocene Closure of the Central American Seaway', *Science*, 348/6231, 2015, pp. 226–9.

48 Programme Tagging of Pelagic Predators (Topp), www.gtopp .org; Peter I. Miller, et al., 'Basking Sharks and Oceanographic Fronts: Quantifying Associations in the North-East Atlantic', *Functional Ecology*, 29/8, 2015, pp. 1099–109; Clive R. McMahon, et al., 'Animal Borne Ocean Sensors – AniBOS – An Essential Component of the Global Ocean Observing System', *Frontiers in Marine Science*, 8, 2021, 751840.

49 The movements of sharks, elephant seals and birds provide information about areas we cannot visit. See, for example, the movements of large pelagic predators in the North Pacific, https://gtopp.org

50 Charlotte A. Birkmanis, et al., 'Shark Conservation Hindered by Lack of Habitat Protection', *Global Ecology and Conservation*, 21, 2020, e00862.

Chapter 8: Fading Silhouettes

1 Libyan waters were hardly exploited during the forty years of Gaddafi's Jamahiriya. Exploitation only intensified after the first years of the 2011–13 war that precipitated his fall.

2 See chapter 3, 'Giving Life'.

3 See note 16, p. 219.

4 The glaucous shark, *Squalus glaucus*, is called the 'blue shark' today. The longnose shark, *Squalus cornubicus*, is the porbeagle, *Lamna nasus*.

5 De Serres, *Des causes des migrations des divers animaux*.

6 Sarano and Granzotto, *Méditerranée*.

7 Prof. M.N. Bradaï, personal communication, during the preparation of the film *Méditerranée, le royaume perdu des requins*, 2013.

8 Béchir Saïdi, et al., 'Shark Pelagic Longline Fishery in the Gulf of Gabes: Inter-Decadal Inspection Reveals Management Needs', *Mediterranean Marine Science*, 20/3, 2019, pp. 532–41.

9 Francesco Ferretti, et al., 'Loss of Large Predatory Sharks from the Mediterranean Sea', *Conservation Biology*, 22/4, 2008, pp. 952–64.

10 Fabrizio Serena and Monica Barone, 'Report on the Meeting of the Working Group on Recreational Fishing', *Newsletter IUCN*, Shark Specialist Group, #1, 2021, p. 60.

11 Or spiny dogfish, *Squalus acanthias*, in danger of extinction today.

12 Anita Conti, *Géants des mers chaudes*, Paris: Payot et Rivages, 1997 [1957].

13 Anita Conti, *La Dame de la mer: Photographe*, Paris: Revue Noire, 1998.

14 Ilse Martinez Candelas, et al., 'Use of Historical Data to Assess Changes in the Vulnerability of Sharks', *Fisheries Research*, 226, 2020, 105526.

15 George Roff, et al., 'Decline of Coastal Apex Shark Populations over the Past Half Century', *Communications Biology*, 1, 2018, 223.

16 Francisco M. Santana, Leonardo M. Feitosa and Rosângela P. Lessa, 'From Plentiful to Critically Endangered: Demographic Evidence of the Artisanal Fisheries Impact on the Smalltail Shark (*Carcharhinus porosus*) from Northern Brazil', *PLoS One*, 15/8, 2020, 0236146.

17 Nathan Pacoureau, et al., 'Half a Century of Global Decline in Oceanic Sharks and Rays', *Nature*, 589, 2021, pp. 567–71.

18 Dulvy et al., 'Overfishing'.

19 Diego Cardeñosa, et al., 'Species Composition of the Largest Shark Fin Retail-Market in Mainland China', *Scientific Reports*, 10/1, 2020, 12914.

20 Save Our Seas, https://saveourseas.com/how-many-sharks-are -caught-each-year

21 Celine J.N. Liu, et al., 'Sharks in Hot Soup: DNA Barcoding of

Shark Species Traded in Singapore', *Fisheries Research*, 241/3, 2021, 105994.

22 Romain Chabrol, *Le prix hideux de la beauté: Une enquête sur le marché de l'huile de foie de requin profond*, Paris: Bloom Association, 2012.

23 www.bloomassociation.org/la-belle-et-la-bete-etude-exclusive -du-requin-dans-nos-cremes-de-beaute

24 Pedro C. Quero-Jiménez, et al., 'Local Cuban Bentonite Clay as Potential Low-Cost Adsorbent for Shark Liver Oil Pool Purification', *Journal of Pharmacy and Pharmacognosy Research*, 9/4, 2021, pp. 525–36.

25 Devita Horax, A.A.G.P. Wiraguna and Gde Ngurah Indraguna Pinatih, 'Topical Administration of Deep Sea Shark Liver Oil (DESSLO™) Inhibit Nuclear Factor-Kappa Beta (NF-kB) Expression in Wistar Rats (*Rattus norvegicus*) Skin Exposed to Ultraviolet-B', *Indonesian Journal of Anti-Aging Medicine*, 4/1, 2020, pp. 5–7.

26 Feily S. Moelyono and Wimpie Pangkahila, 'Administration of Deep Sea Shark Liver Oil Reduce Malondialdehid (Mda) Levels on Male Wistar Rats Exposed to Cigarette Smoke', *Indonesian Journal of Anti-Aging Medicine*, 2/2, 2018, pp. 28–31.

27 Ahmed S. Al Hatrooshi, Valentine C. Eze and Adam P. Harvey, 'Production of Biodiesel from Waste Shark Liver Oil for Biofuel Applications', *Renewable Energy*, 145, 2020, pp. 99–105.

28 See chapter 3, 'Giving Life'.

29 For more information, see the enquiry by the Bloom Association: 'Le poisson dans la restauration scolaire: Nos enfants mangent-ils des espèces menacées?', http://bloomassociation.org/down load/Rapport%20Long%20Cantine%20FR.pdf

30 Terence I. Walker, et al., 'Galeorhinus Galeus-Tope. The IUCN Red List of Threatened Species 2020', Global Shark Trends Project, 2020. DOI: 10.2305/IUCN.UK.2020-2.RLTS.T3 9352A2907336.en

31 James Gelsleichter, et al., 'Elevated Accumulation of the Toxic Metal Mercury in the Critically Endangered Oceanic Whitetip

Shark *Carcharhinus longimanus* from the Northwestern Atlantic Ocean', *Endangered Species Research*, 43, 2020, pp. 267–79.

32 Sang W. Kim, et al., 'Heavy Metal Accumulation in and Food Safety of Shark Meat from Jeju Island, Republic of Korea', *PLoS One*, 14/3, 2019, 0212410; Hitomi Kazama, et al., 'Mercury Concentrations in the Tissues of Blue Shark (*Prionace glauca*) from Sagami Bay and Cephalopods from East China Sea', *Environmental Pollution*, 266/1, 2020, 115192; Liesbeth Weijs, et al., 'Bioaccumulation of Organo-halogenated Compounds in Sharks and Rays from the Southeastern USA', *Environmental Research*, 137/3, 2015, pp. 199–207.

33 Francesca Pancaldi, et al., 'Concentrations of Silver, Chrome, Manganese and Nickel in Two Stranded Whale Sharks (*Rhincodon typus*) from the Gulf of California', *Bulletin of Environmental Contamination and Toxicology*, 107/5, 2021, 827–32; Rubén D. Castro-Rendón, et al., 'Mercury and Cadmium Concentrations in Muscle Tissue of the Blue Shark (*Prionace glauca*) in the Central Eastern Pacific Ocean', *Biological Trace Element Research*, 200/7, 2022, 3400–11.

34 Natascha Wosnick, et al., 'Negative Metal Bioaccumulation Impacts on Systemic Shark Health and Homeostatic Balance', *Marine Pollution Bulletin*, 168, 2021, 112398.

35 Thomas W. Clarkson, Laszlo Magos and Gary J. Myers, 'The Toxicology of Mercury: Current Exposures and Clinical Manifestations', *New England Journal of Medicine*, 349/18, 2003, pp. 1731–7.

36 Moustafa M. Zeitoun and El-Sayed E. Mehana, 'Impact of Water Pollution with Heavy Metals on Fish Health: Overview and Updates', *Global Veterinaria*, 12/2, 2014, pp. 219–31.

37 Kristian J. Parton, 'Elasmobranch (Sharks and Rays) Interaction with Plastic Pollution from Global and Local Perspectives, via Entanglement within Anthropogenic Debris and Synthetic Fibre Ingestion', PhD thesis, University of Exeter, 2019.

38 Kristian J. Parton, et al., 'Investigating the Presence of

Microplastics in Demersal Sharks of the North-East Atlantic', *Scientific Reports*, 10, 2020, 12204.

39 Elitza S. Germanov, et al., 'Microplastics on the Menu: Plastics Pollute Indonesian Manta Ray and Whale Shark Feeding Grounds', *Frontiers in Marine Science*, 6, 2019, 00679.

40 Kristian J. Parton, Tamara S. Galloway and Brendan J. Godley, 'Global Review of Shark and Ray Entanglement in Anthropogenic Marine Debris', *Endangered Species Research*, 39, 2019, pp. 173–90.

41 Juerg Brunnschweiler, Charlie Huveneers and Joanna Borucinska, 'Multi-year Growth Progression of a Neoplastic Lesion on a Bull Shark (*Carcharhinus leucas*)', *Matters Select*, 3/10, 2017.

42 Jessie Walton, 'The Effects of Cyclic Thermal Stress on Oviparous Shark Development', PhD thesis, University of Sudbury, Ontario, 2020.

43 Steven T. Kessel, et al., 'Predictable Temperature-Regulated Residency, Movement and Migration in a Large, Highly Mobile Marine Predator (*Negaprion brevirostris*)', *Marine Ecology Progress Series*, 514, 2014, pp. 175–90.

44 Charles W. Bangley, et al., 'Increased Abundance and Nursery Habitat Use of the Bull Shark (*Carcharhinus leucas*) in Response to a Changing Environment in a Warm-Temperate Estuary', *Scientific Reports*, 8, 2018, 6018.

45 Steven Surina, personal communication.

46 Florian Hoarau, et al., 'Age and Growth of the Bull Shark (*Carcharhinus leucas*) around Réunion Island, South West Indian Ocean', *Journal of Fish Biology*, 99/3, 2021, pp. 1087–99.

47 Alice Manuzzi, et al., 'Retrospective Genomics Suggests the Disappearance of a Tiger Shark (*Galeocerdo ccc*) Population off South Eastern Australia', *Research Square*, 2021.

Chapter 9: The Confrontation

1 G. Grandière, in Patrick and Sophie Durville and Thierry Mulochau, *Comprendre la crise requins à la Réunion*, St Paul: Éditions du Cyclone, 2016, p. 58.

2 Grand requin blanc pêché à la Réunion, 15 October 2015, https:// www.youtube.com/watch?v=EaD8PITTaso

3 F. Sarano and S. Mabile, 'Aux surfeurs et à ceux qui veulent éliminer les requins', *Libération*, 27 August 2012; S. Ribes-Beaudemoulin (pp. 102–3), K. Pothin (pp. 96–7), and J.-F. Nativel (p. 85), in Durville and Mulochau, *Comprendre la crise requins à la Réunion*.

4 C. Mulquin, in Durville and Mulochau, *Comprendre la crise requins à la Réunion*, p. 83.

5 M. Soria, *Comprendre la crise requins à la Réunion*, pp. 110–11; M. Soria, 'Bilan de l'analyse des données de marquage collectées du mois de décembre 2011 au mois de septembre 2013 dans le cadre du programme charc', in M. Soria, ed., Saint-Denis de la Réunion: IRD, 2014, p. 31; Rapport scientifique final du programme charc (Connaissances de l'écologie et de l'habitat de deux espèces de requins côtiers sur la côte ouest de la Réunion), Étude du comportement des requins-bouledogues (*Carcharhinus leucas*) et tigres (*Galeocerdo cuvier*) à la Réunion, IRD, 2015.

6 Nativel, in Durville and Mulochau, *Comprendre la crise requins à la Réunion*, p. 85.

7 He had already tried a first unsuccessful mediation and proposed dives to count the sharks in Saint Leu on 24 September 2013.

8 Le Dodo palmé diving club.

9 See publication of 27 February 2020, www.ipreunion.com/requins/reportage/2020/02/27/opr-bons-d-achat-requins-opr-bons-d-achat-requins115001.html [link no longer available].

10 Press report of 17 April 2021, https://la1ere.francetvinfo.fr/reunion/saint-andre/saint-andre-un-requin-bouledogue-de-2-50m-peche-a-champ-borne-986440.html

11 A total of 166 offences were recorded by the Scientific Council of the marine reserve, offences committed by fishermen from the Shark Safety Centre, an official body chaired by the sub-prefect. Following this repeated poaching, which has been denounced

many times, the Scientific Committee published a self-referral on 19 May 2021, which was picked up by the newspaper *Le Monde* on 14 June 2021, and then sent a letter to the local and national authorities on 19 June.

12 Marianne Énault, 'Pêche aux requins en eaux troubles', *Journal du Dimanche*, 8 August 2021.

13 Association Saving Our Sharks, www.savingoursharks.org

14 Mallory G. McKeon and Joshua Drew, 'Community Dynamics in Fijian Coral Reef Fish Communities Vary with Conservation and Shark-Based Tourism', *Pacific Conservation Biology*, 25/4, 2018, pp. 363–9.

15 International Shark Attack File, https://www.floridamuseum.ufl.edu/shark-attacks/

16 Jacqueline Goy and Robert Calcagno, *Méduses: À la conquête des océans*, Monaco: Éditions du Rocher, 2014.

17 Bodysurfing: riding the wave without a board, with a bare body.

18 Worldwide, 2.5 million divers make an average of sixteen million dives each year. In comparison, there are twenty-three million surfers, according to the International Surfing Association.

19 Attack of a grey shark on Julien Leblond, www.facebook.com/watch/?v=499900274066209

20 Donald R. Nelson, 'Aggression in Sharks: Is the Grey Reef Shark Different?', *Oceanus*, 24/4, 1981–2; Johnson, *Sharks of Polynesia*.

21 Diving equipment that recycles air and therefore does not release the air exhaled by the diver back into the water.

22 Ghislain Bardout, personal communication, 2021.

23 Johann Mourier, personal communication.

24 Attack of a tiger shark on Céline Lefebvre, diving instructor, 31 December 2020, www.youtube.com/watch?v=eFnO-0GSiV0&feature=share&ab_channel=SharkIslandTV; Céline Lefebvre, personal communication, 2021.

25 Eric Clua, 'Fatal Shark Attacks on Humans: Rather an Individual

Behavioral Problem than a Collective Ecological Issue', MS for *Behavioral Ecology*, 2015. Clua and Linnell, 'Individual Shark Profiling'.

26 The Malibu Artist, video of a great white shark among swimmers, www.youtube.com/watch?app=desktop&v=Ile5NS7ucec& feature=share&fbclid=wAR3EKTud-bKFrCE1uB1NQ6QBtAmj AnWs3PtFqJ_WpBmHITnHvWYiBlRIKLE

27 Christopher Neff, 'Australian Beach Safety and the Politics of Shark Attacks', *Coastal Management*, 40/1, 2012, pp. 88–116.

28 Ibid.

29 Geremy Cliff and Sheldon F.J. Dudley, 'Reducing the Environmental Impact of Shark-Control Programs: A Case Study from KwaZulu-Natal, South Africa', *Marine and Freshwater Research*, 62, 2011, pp. 700–9.

30 Bradley M. Wetherbee, Christopher G. Lowe and Gerald L. Crow, 'A Review of Shark Control in Hawaii with Recommendations for Future Research', *Pacific Science*, 48/2, 1994, pp. 95–115.

31 Prefectural review as of 31 August 2021.

32 'Arrêtons de prendre des risques', *Le Quotidien de la Réunion*, 28 August 2006.

33 Éric Clua, Programme Progenir pour le profilage génétique individuel des requins, www.thesharkprofile.com

34 Clua and Linnell, 'Individual Shark Profiling'.

35 Ibid. https://thesharkprofile.com; Eric Clua and Dennis Reid, 'Contribution of Forensic Analysis to Shark Profiling Following Fatal Attacks on Humans', in Kamil Hakan Dogan, ed., *Post Mortem Examination and Autopsy: Current Issues from Death to Laboratory Analysis*, Rijeka, Croatia: IntechOpen, 2018, chap. 5.

36 Nativel, in Durville and Mulochau, *Comprendre la crise requins à la Réunion*, p. 85.

37 Daryl P. McPhee, et al., 'A Comparison of Alternative Systems to Catch and Kill for Mitigating Unprovoked Shark Bite on Bathers

or Surfers at Ocean Beaches', *Ocean and Costal Management*, 201, 2021, 105492.

38 Durville and Mulochau, *Comprendre la crise requins à la Réunion*, p. 76. Author's emphasis.

39 Ibid., p. 85.

40 Ibid., p. 15.

41 Jacques-Yves Cousteau and Jean-Michel Cousteau, *La Machine à remonter le temps: Papouasie-Nouvelle-Guinée 1*, doc., 52 min, 1989, https://www.youtube.com/watch?v=Vjw0yv10CmI

42 Abrantes and Barnett, 'Intrapopulation Variations in Diet and Habitat Use'.

43 A tourist activity that consists of diving to observe reefs at sites where they are concentrated more or less artificially.

44 Kirin Apps, Kay Dimmock and Charlie Huveneers, 'Turning Wildlife Experiences into Conservation Action: Can White Shark Cage-Dive Tourism Influence Conservation Behavior?', *Marine Policy*, 88, 2018, pp. 108–15; Serena Lucrezi, Filippo Bargnesi and Francois Burman, '"I Would Die to See One": A Study to Evaluate Safety Knowledge, Attitude, and Behavior among Shark Scuba Divers', *Tourism in Marine Environments*, 15/3–4, 2020, pp. 127–58.

45 Sarah R. Sutcliffe and Michele L. Barnes, 'The Role of Shark Ecotourism in Conservation Behavior: Evidence from Hawaii', *Marine Policy*, 97, 2018, pp. 27–33.

46 Céline Lefebvre, personal communication.

47 Fin for a Fin Association, www.facebook.com/finforafin

48 Founder of the Malpelo Foundation and other marine ecosystems. She is the initiator and first director of the Malpelo Fauna and Flora Sanctuary, which she had listed as a World Heritage Site in 2006.

49 https://whc.unesco.org/fr/list/1216; www.fundacionmalpelo.org; https://whitleyaward.org/winners/shark-conservation-malpelo-world-heritage-site-colombia

50 Sandra Bessudo, personal communication.

51 Shark Education, www.sharkeducation.com

52 Steven Surina, personal communication.
53 Rob Stewart, *Sharkwater*, DreamWorks SKG/Universal Pictures, doc., 90 min, 2008. www.sharkwater.com
54 Sea Shepherd, https://seashepherd.org/
55 Longitude 181, www.longitude181.org
56 A term referring to various biodiversity initiatives (e.g., https://www.facebook.com/coloca.terres.1), incorporating 'Earth' (*Terre*), but also homophonic with *colocataire*, 'flatmate' – translator's note.
57 L 181, www.longitude181.org/wp-content/uploads/2016/07/historique_campagne_polynesie.pdf
58 Laurent Ballesta, *700 requins dans la nuit.*
59 Christine A. Ward-Paige, et al., 'Recovery Potential and Conservation Options for Elasmobranchs', *Journal of Fish Biology*, 80/5, 2012, pp. 1844–69.

Chapter 10: Reconciliation

1 One can read *fils* either way: threads, elements of a fabric, or sons, children of the living.
2 'À quoi servent les Gitans de Perpignan', *Le Petit Journal catalan*, 20 August 2015.
3 It is estimated that the number of rays and sharks has been reduced by 70 per cent over the last fifty years.
4 Marine creatures are so far from our world that when a fish was seen jumping out of the water off Ploemeur (Morbihan), the Kaolins beach was closed and swimming forbidden! (*Ouest-France*, 1 September 2021). In the Mediterranean, on 21 July 2020, in the late afternoon, the fire brigade was called to the Fighière beach in Villeneuve-Loubet to evacuate the bathers and ban swimming, after a so-called 'cow-nosed' ray (or 'mourine', totally harmless) had been spotted! (*Nice-Matin*, 21 July 2020). Elsewhere, a small ballista inconveniencing bathers made the front page of *La Dépêche* on 30 July 2020.
5 Nathan Pacoureau, et al., 'Half a Century of Global Decline in Oceanic Sharks and Ray', *Nature*, 589, 2021, pp. 567–71.

6 Collectif, *Le Petit Prince: Dessine-moi ta planète*, Arles: Deyrolle/ Actes Sud, 2021, pp. 50–3.
7 Gary, 'Letter to an Elephant'.
8 Roger Allers and Robert Minkoff, *The Lion King*, Disney Studios, 1994.
9 Véronique Sarano and François Sarano, *Libye*, Paris: La Manufacture, 2001.
10 Jacques Sarano, *L'Homme double: Dualité et duplicité*, Blois: Éditions de l'Épi, 1979.
11 Morizot, *On the Animal Trail*.
12 Baptiste Morizot, *Ways of Being Alive*, trans. A. Brown, Cambridge: Polity, pp. 232–41.
13 Schaub, *François Sarano*, p. 173.
14 Ibid., p. 191.
15 Perrin and Cluzaud, *Oceans*; Stéphane Durand and François Sarano, *Océans*, Paris: Seuil, 2009, pp. 44–57.
16 Sarano and Granzotto, *Méditerranée*.
17 Morizot, *Ways of Being Alive*, p. 24.

List of Illustrations

Fig. 1. The final shark cage dive in the Andaman Islands.
© Marion Sarano   10–11

Fig. 2. Escaped convicts on a raft surrounded by sharks.
Le Petit Journal Illustré, 20 May 1906. © Véronique
et François Sarano   20

Fig. 3. *Le Dauphiné Libéré*, 18 July 1958. © Véronique and
François Sarano   24

Fig. 4. Fossil of the sun ray and the angelshark. © Marion
Sarano   32

Fig. 5. Skate (ray) and shark. © Marion Sarano   33

Fig. 6. Geological time and vertebrate phylogeny.
© Marion Sarano   35

Fig. 7. Evolution of the world's oceans. © Marion Sarano   38–9

Fig. 8. Egg of an elephant fish (chimaera). © Marion Sarano   41

Fig. 9. First encounter with the chimaera. © Marion Sarano   42

Fig. 10. General morphology of modern sharks. © Marion
Sarano   45

Fig. 11. *Helicoprion.* © Marion Sarano   47

Fig. 12. Fossil tooth of *Otodus megalodon.* © Marion Sarano   52

Fig. 13. Whale shark eye and dwarf lanternshark. © Marion
Sarano   54–5

Fig. 14. Key to the identification of the major orders of
         sharks. © Marion Sarano   56
Fig. 15. School of hammerhead sharks. © Marion Sarano   58–9
Fig. 16. Schematic representation of the range of shark
         senses. © Marion Sarano   78
Fig. 17. Pores of the ampullae of Lorenzini distributed over
         the head of a great white shark. © Véronique and
         François Sarano   82
Fig. 18. Privacy sphere of a great white shark. © Pascal
         Kobeh/Galatée Films   84–5
Fig. 19. Ana, a female bull shark. © Véronique and François
         Sarano   89
Fig. 20. The gaping mouth of the manta ray. © Véronique
         and François Sarano   93
Fig. 21. A whale shark devouring the fish who had taken
         refuge next to it. © Véronique and François Sarano   111
Fig. 22. Parasitic copepods. © Véronique and François Sarano   124
Fig. 23. Parasitic remora fish. © Véronique and François Sarano   124
Fig. 24. Thermohaline circulation of the world ocean.
         © Marion Sarano   144–5
Fig. 25. First visual identification of 'El monstro'.
         © Véronique and François Sarano   151
Fig. 26. Unloading of grey sharks at Zarzis, Tunisia.
         © Véronique and François Sarano   155
Fig. 27. Blue sharks caught during a 'sport' fishing
         competition. © Véronique and François Sarano   158
Fig. 28. The author sketching in Guadalupe. © Gérard Soury   177
Fig. 29. Contortions of the great white shark. © Véronique
         and François Sarano   178
Fig. 30. Lady Mystery and the author, shoulder to shoulder.
         © Pascal Kobeh/Galatée Films   208–9
Fig. 31. Lady Mystery carries the author on her pectoral
         fin. © Pascal Kobeh/Galatée Films   210

Acknowledgements

Many thanks to:

Véronique, Marion, Maud, Ayaté, Brice; without you who give me life on a daily basis, there would be no book. It would even be more accurate, Véronique, to acknowledge you as co-author.
   Stéphane; this book owes a lot to your delicate proofreading, your encouragement and your friendship.

Anne-Sylvie, Françoise, Jean-Paul; thank you, thank you for your trust and your always wonderful and delicious hospitality.

Marie-Amélie, Vanessa; for having put the p's and q's back in their right places and for having an eagle eye for incomplete bibliographical references.

Sandra Bessudo; your strength of conviction, with a smile, constantly motivates us to protect our wild cousins of the oceans.

Pascal Kobeh; photographer and accomplice, who immortalized our encounters with Lady Mystery during the shooting of the film *Océans*.

Gérard Soury; who knows sharks so well, author of the photo taken during the filming in Guadalupe in 2013.

I would also like to thank Dr Johann Mourier, shark specialist, and Dr Eric Clua, from the École Pratique des Hautes Études, pioneer in the study of shark personality, for their advice and their sharp and judicious proofreading. Céline Lefebvre, Steven Surina, Ghislain Bardout and Laurent Debas for their valuable testimonies.

This book is the result of hundreds of dives with sharks in all seas of the world. All of them were done as a team and are therefore the fruit of the collaboration and friendship of those who were at my side during these unforgettable encounters. In chronological order since the 1970s: my friends from the Cercle Valence Plongée; Fernand Voisin and the crew of the *Petrouchka*; Jacques-Yves Cousteau, Albert Falco and all the crew of the *Calypso*; Christian Pétron, Yves Lefèvre, Jean-Marc Bourg, Luc Hieulle, Pascal Szymanek, in memory of the first encounters with the great whites alongside Andre Hartman; Jacques Perrin, Jacques Cluzaud, Olli Barbé, Patricia Lignières and the crew of the film *Océans*; René Heuzey and Stéphane Granzotto; Didier Manenq and the Phocea Mexico team; Fred Bassemayousse and all my friends from the Longitude 181 association, chaired by Patrice Bureau for the past ten years, who work behind the scenes on a daily basis to offer future generations an ocean richer than the one we know today.

Index

Page numbers in *italics* denote a figure

abandonment, at birth, 67, 96
Abrantes, Kátya Gisela, 97
abyssal sharks, 133
accidents, unprovoked, 173–4,
    178–83, 191
Africa, 160
Alcyone, 132, 136
algae, 30, 65, 106, 116, 150, 168
amino acids, 75
amphibians, 92
'ampullae of Lorenzini', 81, *82*, 113,
    221n.21
Ana (female bull shark), 88–90, *89*
Anacharsis, 14, 213n.14
Anaxi Mandra of Miletus, 14,
    213n.16
anchovies, 110, 155
Andaman Islands, 4, 5–6, *10–11*
Anelasma squalicola (barnacle), 123
angelfish, 128, 192
angelshark (*Squatina squatina*),
    31, *32*
angular roughshark (*Oxynotus
    centrina*), 14

animated films, 26–7
Anthropocene Epoch, 66–7,
    219n.20
anthropogenic pollution, 164,
    225n.2
Antoinette (great white shark),
    136–7
apex-predator shark, 115
aquariums, 108, 198
Aran Islands, 51
Argentina, 101
Ari, Csilla, 99
Aristotle, *History of Animals*, 13, 34
Arthur (sperm whale), 99
asexual reproduction, 68
Asia, 53
Atlantic Dawn, 107
Atlantic Ocean, 51, 140, 141–2, 166
attacks, shark, 20–1, 173–4
Australia, 136–7, 140, 141, 142, 183
Ayaté (female bull shark), 90

Baby Shark (YouTube video), 28
bacterial infections, 122–3

Bahamas, 61, 173
Baja California, 139
Ballesta, Laurent, 113, 118, 121, 196
bamboo shark (*Chiloscyllium punctatum*), 166
banana fusiliers (*Pterocaesio pisang*), 9
Bardout, Ghislain, 175–6
barnacle 123, 150
barracuda (*Sphyraena barracuda*), 61, 192
basking shark (*Cetorhinus maximus*), 53, 67, 142–3
Bassemayousse, Fred, 160
bats, 87
Bear Gulch fossil site, Montana, 44
Belantsea montana (extinct fish), 44
Belon, Pierre, *Histoire naturelle des estranges poissons marins*, 14
Benchley, Peter, *Jaws*, 25
Bering Strait, 135
Bessudo, Sandra, vii, 149, 152, 192–3, 246n.48
bigeye trevally (*Caranx sexfasciatus*), 109, 129
'biofuel', 163
birds, 130, 201
black jack fish (*Caranx lugubris*), 117
black market, fin, 161–2
blacknose shark (*Carcharhinus acronotus*), 142
blacktip reef shark (*Carcharhinus melanopterus*), 95, 97
blacktip shark (*Carcharhinus limbatus*), 83
blotched fantail rays (*Taeniura meyeni*), 72
blue shark (*Prionace glauca*), 13, 37, 57, 123, 156–7, *158*, 161, 238n.4
Blue Water, White Death (Lipscomb and Gimbel), 23–4
Blue Wolf (*Börtea-Chino*), 16

bluefin (*C. melampygus*), 71
bluefish (*Pomatomus saltatrix*), 101
bluestreak cleaner wrasse (*Labroides dimidiatus*), 127, 128
bluestripe snapper (*Lutjanus kasmira*), 168
bluntnose sixgill shark (*Hexanchus griseus*), 46, 48, 49, 66, 76, 80, 131
bonito tuna, 110
bonnethead shark (*Sphyrna tiburo*), 83
bony fish, 34, 63, 68
boobies, 121
Bora Bora reef, French Polynesia, 174
bottlenose dolphin (*Tursiops truncatus*), 102
Bourail, New Caledonia, 179
Bourg, Jean-Marc, 149
bowmouth guitarfish (*Rhina ancylostoma*), 31
broadnose sevengill shark (*Notorynchus cepedianus*), 97
Bruce, Barry, 136, 137
Bruder, Charles, 21
bull shark (*Carcharhinus leucas*), 166–7, 169–70, 171–3
  accidents and, 179–80
  characteristics, 222n.2
  Coles on, 101
  feeding sites, 188
  females, 88–90, *89*, 95, 172
  gaze of, 79
  gestation period, 66
  gill slits, 31
  injured/killed, 165, 184
  sphere of intimacy, 84
Bureau, Patrice, 169, 243n.7
Burghardt, Gordon, 92
butterfly fish (*Johnrandallia nigrirostris*), 126, 127, 128, 129
Byrnes, Evan, 95

cages, shark, 7–8, *10–11*
calving, 60–1, 62, 66–7, 114, 139–40, 171
Calypso
  cameras, 72, 220n.1
  Cousteau and, 6–7, 23
  Cuba, 110–12
  Indian Ocean, 4, 5
  New Zealand, 36
  Papua New Guinea, 132, 188
  Raine Island, 29, 121–2
  shark cage, 8
cancers, 122–3, 165
Canterbury Bay, New Zealand, 37
Cape Agulhas, 131, 230n.48
Cape gannet (*Morus capensis*), 104–5, 120
Cape La Houssaye, Réunion Island, 168–9
Cape Lookout, North Carolina, 101
Cape of Good Hope, 130, 131
Cape penguin (*Spheniscus demersus*), 130–1
Cape Verde Islands, 6–7, 160
Carcharhiniformes, 80
Caribbean Sea, 88–90, 160
carpet sharks, 31
'Carta Marina' (Magnus), 14
cartilage, shark, 122–3
cartilaginous skeletons, 34, 36, 37, 43, 44
Celorio, Apolinario, 193
cephalic horns, 93, *93*
cetaceans
  corpses, 133
  feeding, 104–5, 120, 131, 132
  juveniles, 67, 96, 98, 114
  reproduction, 13
  social network, 95
character traits, 94–5
charge, intimidation, 176–7, *177–8*
chimaeras, 36, 37–43, *41–2*, 44
Chinese trumpetfish (*Aulostomus chinensis*), 168

chondrichthyans, 34, 36–7, 44, 57, 81
cirripedean crustaceans, 123
Cladoselache (extinct shark-like fish), 43–4
classes, shark, 34, *56*, 57
cleaner fish, 127–9
climate change, 115, 147, 166
Clua, Eric, 25, 90, 97, 179, 185
Cluzaud, Jacques, 130
Cocos Islands marine reserve, Costa Rica, 129, 193
Coenen, Adriaen, *Whale Book*, 14–15
Coiba Island, Panama, 193
cold-blooded animals, 25, 51, 91, 92–3, 215n.36
Coles, Russell J., 101
coloca-Terres, 195, 207, 247n.56
Colombia, vi, vii, 126, 162, 192–3
colossal whale shark, 63
common dogfish (*Scyliorhinus canicula*), 45
common sawshark (*Pristiophorus cirratus*), 31
Commonwealth Scientific and Industrial Research Organisation (CSIRO), 137
communication, 101, 148
Comprendre la crise requins à la Réunion (Durville and Mulochau), 186
conger eels, 64, 159
consciousness, animal, 99–100
Conti, Anita, 159–60
convicts, sharks and, *20*
cookiecutter (*Isistius brasiliensis*), 57
copepods, 123, *124*, 231n.59
Copley, John Singleton, *Watson and the Shark*, 16
copper shark (*Carcharhinus brachyurus*), 105, 117, 120
coral, 64, 68, 71, 126, 128

coral reefs, 8–9, 67, 76, 113, 120,
    128, 168, 173
Coral Sea, 29–30, 108–10, 121–2
coral sharks, 71, 76, 109, 113–14,
    117, 120, 129
Costa Rica, vii, 129, 193
Cousteau, Jacques, 4, 6–8, 12, 95
    Silent World, 5, 22–3
Cousteau, Jean-Michel, 136–7
Cousteau, Philippe, 23
cownose stingray, 116, 247n.4
creams, beauty, 162–3
crinoids, pedunculated
    (*Gymnocrinus richeri*), 50
crustaceans, 83, 115–16, 123, 125,
    133
Ctenacanthus (extinct shark), 43
Cuba, 110–12, *111*
culling, 169, 183–7
currents, 48–9, 77, 105–6, 141–2,
    144, 146, 149–52, 225n.3,
    229n.36

Damocles serratus (extinct shark),
    46
damselfish (*Dascyllus albisella*), 148
Dangerous Reef, Australia, 136–7
Dauphiné libéré, 23, *24*
De Piscibus Marinis (Rondelet), 15
de Serres, Marcel, 16–17, 136, 141,
    155–6
Debas, Laurent, 227n.21
Deep Blue (great white shark),
    139–40
Deep Blue Sea (1999 film), 26, 108
Deep Ocean Odyssey programme,
    131
Deloire, Michel, 108
demigods, 19
denticles, 34, 46, 124
Dérand, Didier, 194–5
Deval, André, 23
Devonian period, 44
diet, 40, 43, 97–8, 106–8, 117

digestive bacteria, 125
diversification, natural, 27, 57–8,
    67–8, 120, 148, 202
dogfish (*Squalus acanthias*), 13,
    64–5, 66, 131, 159, 163
dolphin (*Delphinus capensis*), 92,
    101, 104–5, 107, 120–1
dugong (*Dugong dugon*), 114
Dumas, Frédéric 'Didi', 95
Dunkleosteus (extinct fish), 43–4
Dunn, Joseph, 21
Durville, Patrick and Sophie,
    *Comprendre la crise requins à
    la Réunion*, 186
dusk, diving at, 9
dusky dolphin (*Lagenorhynchus
    obscurus*), 101
dusky smoothhound (*Mustelus
    canis*), 79, 117
dwarflantern shark (*Etmopterus
    perryi*), 53, *55*, 124
Dyer Island, Cape of Good Hope,
    130

eagle ray, 31, 72
Easter Island, 74
ecological amnesia, 64, 155,
    219n.16
ecosystems, 6, 97–8, 104–33, 173,
    187, 197–8, 203, 213n.3
ectopic (definition), 215n.2
Ecuador, vii
Edestus giganteus (extinct shark),
    44, 46
eggs, 63–6
    fish, 116
    green turtle, 29, 122
    shark, 37, *41*, 166, 229n.36
El Arrecife, Malpelo Island, 126
'*El monstro*' *see* smalltooth sand
    tiger shark
El Niño, 106, 225n.3
elasmobranchs, 92
electrosensing, 73, 81, 83

elephant fish (*Callorhinchus milii*), 36, 37–43, *41–2*, 44
elephant seal (*Mirounga angustirostris*), 132
Elie Monnier, 7
elimination, shark, 68–9, 118, 119, 121–2, 166–7, 169, 183–7, 200
Eliot (sperm whale), 99
epaulette shark (*Hemiscyllium ocellatum*), 30
Eqalussuaq, *see* Greenland shark
Eratosthenes, 14, 214n.17
exaptation, error and, 148–9, 237n.43
exploitation, 5, 155, 161–4, 195
extinction, 43–4, 46, *47*, 49, 57, 62, 63
eyelids, 80
eyes, *54*, 79–80, 123

Fakarava, French Polynesia, 113, 118, 121, 176
familiarity, of shark to all, 12
Farallon Islands, California, 132
fecundity, low, 66, 155, 167
feeding, 106–8, 112–18
    bull shark, 188
    on corpses, 133
    great white shark, 107
    grey reef, 108, 113–14, 120
    oceanic manta rays, 93–4, *93*
    sites, 188–9, 227n.18, 229n.36
    'sphere of intimacy', 85
    tiger shark, 29, 122
Fenrir (wolf), 16
Ferrucci, Aldo, 206
fertilization, internal/external, 61–3
Fiji, 165, 173, 188
fils (threads), 198, 247n.1
Fin for a Fin (surfers' association), 191–2
Finding Nemo (2003 film), 26
finning, shark, vi, 161–2, 194

fins
    caudal, 227n.17
    paired, 36–7
    pectoral fin joint, 44
fish, 34, 61–2, 105–6, 109–18, 201, 232n.67, 247n.4
Fisher, Watson Stanley, 21
fisheries, 160, 163–4, 183
fishermen, 13, 188, 193, 243n.11
fishing, cooperative, 93–4
fishing, recreational, 157–9, 182
'fission–fusion', 94–5
Flaherty, Robert, *Man of Aran*, 53
flatback turtle (*Natator depressus*), 114–15
'food chain', marine, 116–17, 229n.36
'food pyramid', 116, 228n.33
fossils, *32*, 43, 44, 46, *47*, *52*, 62, 142
fox shark, *see* thresher sharks
France, 16–17, 136, 247n.4
French Polynesia, 113, 118, 121, 174, 195–6
frilled shark (*Chlamydoselachus anguineus*), 50, 66
Frizzia (bull shark), 166
fur seal (*Arctocephalus pusillus*), 130, 131

Galapagos archipelago, 152, 193
Galeus and *Poroderma*, 132
Gary, Romain, 36
gene redundancy, 122
genetic mixing, 63
gestation period, 66, 139–40
Gévaudan, Beast of, 16–17, 19
ghost shark (*Callorhinchus milii*), *see* elephant fish
Giacoletto, Yvan, 108
giant freshwater whipray (*Urogymnus polylepis*), 57
giant sperm whale, 57–8
Giant squid (*Architeuthis*), 139

Giant trevallies (*Caranx ignobilis*), 71

gills, 37, 232n.67

slits, 31, *33*, 34, 46, 49, 50

Gimbel, Peter, *Blue Water, White Death*, 23, 25

glaucous shark (*Squalus glaucus*), 156, 238n.4

'globe-swimmers', 143

gnathostomes, 36, 229n.36

goldfish, 34

Grand Bahama, USA, 60–1

Granzotto, Stéphane, 156, 206

Great Barrier Reef, 67, 183

great hammerhead (*Sphyrna mokarran*), 173

great white shark (*Carcharodon carcharias*), 136–41, 205–7
　　Antoinette, 136–7
　　charge of, 176–7, *177–8*
　　Deep Blue, 139–40
　　diving with, 25
　　evolution of, 13, 51
　　expedition, 130–1
　　feeding, 107
　　fictional, 27
　　fishing for, 157
　　Guadalupe Island, 190
　　juvenile, 97, 114
　　Lady Kathy, 138, 205
　　Lady Mystery, 176, 198, 205–7, *208–10*, 209, 211
　　'lamia', 15
　　length, 215n.33
　　low fecundity, 167
　　Nicole, 137
　　Nukumi, 140, 235n.22
　　parasites on, 123, 124, *124*
　　personality, 90
　　reproduction, 156
　　sense organs, 76, 80, *82*
　　South Africa, 173
　　'sphere of intimacy', *84–5*
　　Squalus carcharias, 156, 234n.8

tracking, 235n.22
Trigger, 176, 205
Ugly, 176

Greek mythology, 12

green humphead parrotfish (*Bolbometopon muricatum*), 109

green turtle (*Chelonia mydas*), 29, 114, 121–2

Greenland shark (*Somniosus microcephalus*), 67, 123, 134–5

Greenland, southern, 140

grey bamboo shark (*Chiloscyllium griseum*), 96

grey reef shark (*Carcharhinus amblyrhynchos*)
　　Andaman Islands, 5
　　behaviour, 175–6
　　cleaner fish, 127–8
　　Coral Sea, 109
　　feeding, 108, 113–14, 120
　　females, 6
　　injured, 6
　　Zarzis, *155*, 157

groupers (*Epinephelus polyphekadion*), 113, 118, 121, 122, 227n.21

Guadalupe Island, 107, 128, 137–40, 176–7, *177*, 190, 205–6

Guangzhou, China, 162

Gulf of Gabes, 156

Gulf of Mexico, 135, 140, 160

Guttridge, Tristan, 102

hagfish, 36

hammerhead sharks, vii, 14, *58–9*, 117, 126–7, 129, 161, 193

Handerson, Georges, 195–6

Hanifaru Bay, Maldives, 94

Hartman, Andre, 25

hatchetfish (*Argyropelecus aculeatus*), 49

Hawaii, 135, 139, 183

head, shark, 71–87

hearing, 77–9
Helicoprion (extinct shark-like fish), 46, *47*
Hemingway, Ernest, *The Old Man and the Sea*, 22, 74–5
hermaphroditism, 68
Hermit Islands, Papua New Guinea, 188
Heuzey, René, 157, *177*, 206
Hinemoana, 15
Histoire naturelle des estranges poissons marins (Belon), 14
History of Animals (Aristotle), 13, 34
hogfish (*Bodianus diplotaenia*), 126, 128
Holdenius (extinct fish), 43
Homer, *The Odyssey*, 48
Hong Kong, 162
houndshark (*Mustelus asterias*), 163
Hugo, Victor, *Les Travailleurs de la mer*, 17
Humane Society International, 183
Hurricane Kate (1985), 110

identity, 96–7
Indian Ocean, 4, 5, 104, 137
Indianapolis, USS, 21
Institut de recherche et de développement (IRD), 169
International Shark Attack File (ISAF), 173
International Union for Conservation of Nature (IUCN), 159, 161, 163–4
intimidation charge, 176–7, *177–8*
Invisible Bank, Andaman Islands, 5
Iroquois, 16

Jacquard, Albert, 204
Japan, 21, 165
jaws, 36–7
Jaws (1975 film), 25, 26, 122–3

Jaws (Benchley), 25
jellyfish, 49
Jenny (bull shark), 172
Johnson, Dr Richard, 76, 175
Julien (injured diver), 175
Jurassic Park (1993 film), 48
juveniles, 97, 108, 114, 140

Ka'ahupahau, 15
Kais II, 156–7
Kamel (fisherman), 156, 157
Kamohoalii, 15
Kane-i-kokala, 15
Kawariki, 15
killer whale, 101, 114, 131–2
Kobeh, Pascal, 138
kraken, 15
KwaZulu-Natal, South Africa, 183

La Couronne, Marseille, 64, 159
Labourgade, Pierre, 120
Lady Kathy (great white), 138, 205
Lady Mystery (great white), 176, 198, 205–7, *208–10*, 209, 211
Lamia, daughter of Poseidon, 12, 15
lamia, 15
lampreys, 36, 124
Lawson Bank, Marquesas Islands, 71–2
Le Neindre, Pierre, 100
leather bass (*Dermatolepis dermatolepis*), 126
Lee Creek Mine site, USA, 51
Lefebvre, Céline, 178–9, 191
Lefèvre, Yves, 149, 150
lemon shark (*Negaprion brevirostris*), 60–1, 95, 98, 102, 166
Libya, 154, 157, 201, 238n.1
Licciardi, Jean, 157
Lipscomb, James, *Blue Water, White Death*, 23, 25
literature, sharks in, 17–19, 22, 25
little skate (*Leucoraja erinacea*), 30

liver oil (squalane), 162–3
lobster, 64, 159
loggerhead sea turtle (*Caretta
caretta*), 168
Longitude 181, 194–5, 247n.55
longnose shark (*Squalus
cornubicus*), 156, 238n.4
Lorenzini cells, 81, 82, 113, 221n.21
lycodes fish (*Pachycara crassiceps*),
133

mackerel, 117, 136, 155–6
magnetic fields, 146–7
Magnus, Olaus, 'Carta Marina', 14
mako shark *see* shortfin mako shark
Malpelo Island, Colombia, 126,
149–52
Malpelo Marine Reserve, 192–3
Malta, 141
Man of Aran (Flaherty), 53
mangrove forests, 60–1
manta ray (*Mobula alfredi*), 91,
93–4, *93*, 99
Maoris, 15
Marco (diver), 150
Marquesas Islands, 71–2, 76
Martine (female bull shark), 90
Materpiscis attenboroughi (extinct
fish), 62
mating, 29, 61–2, 66, 122, 127, 138
maturity, sexual, 60, 66–7, 114, 135,
157, 198
Mauritius, 98
media, 16, 21–3, 181, 185
Mediterranean Sea, 16–17, 49, 64,
141, 142, 156–7, 159
megalodon shark (*Otodus
megalodon*), 43, 50–1, *52*
megamouth shark (*Megachasma
pelagios*), 50
Melanesia, 15
Melville, Herman, *Moby Dick*, 17
Mentawai Sea, 91
mesopredators, 115

metal pollution, 165
Mexico, 31, 171–2
Microbrachius dicki (extinct fish),
63
migration, 81, 136, 143, 166, 185,
193, 230n.48
Miocene Epoch, 142
mirror test, 99, 224n.32
Moby Dick (Melville), 17
Mongolians, 16
monochromatic vision, 80
monsters, 14–15, 17–19
Morizot, Baptiste, 203, 211
morphology, of sharks, 34, *45*, 50
'morphology–physiology–
behaviour' combination, 69
Mourier, Johann, 95, 97, 120, 176
Mulochau, Thierry, *Comprendre la
crise requins à la Réunion*,
186
Myers, Prof Ransom A., 116
mythology, 12, 15–16

Nagel, Thomas, 87
National Marine Nature Reserve,
Réunion Island, 169, 243n.11
natural selection, 68–70, 197
naturalists, early, 13
nautilus (*Nautilus* sp.), 44, 50
Neff, Christopher, 181–2
Nelson, Donald, 175
nets, shark, 182–3
New Caledonia, 50, 67
New Jersey coast, 20–1
New Zealand, 36, 140, 141, *142*
Nicole (great white), 137
ninja lanternshark (*Etmopterus
benchleyi*), 123
'Noah's Ark' concept, 26, 27, 198
Noirot, Didier, 8, 138, 205
North Atlantic, 140, 141
North Carolina, 101, 166
North Horn, Osprey Reef, 108–10
North Pacific Ocean, 139, 238n.49

Notidanoides (extinct shark), 49
Nouvian, Claire, 162
Nukumi (great white), 140, 235n.22
nurse shark (*Ginglymostoma
    cirratum*), 57, 62, 83, 108
nursehound (*Scyliorhinus stellaris*),
    64

Oahu, Hawaii, 139
oceanic manta ray (*Mobula
    birostris*), 31, 53, 92–4
oceanic whitetip shark
    (*Carcharhinus longimanus*),
    6–7, 79, 95, 213n.5
oceans
    evolution of the world's, 38–9, 46,
        141
    twilight zone, 139, 235n.17
Oceans (2009 film), 25, 130, 137,
    176–7, 205–7, *208–10*
octopus, giant, 17–18
The Odyssey (Homer), 48
The Old Man and the Sea
    (Hemingway), 22, 74–5
opinions, changing, 188–9
ornate wobbegong (*Orectolobus
    ornatus*), 30
overfishing, 157
oxygen production, 115, 116

Pacific jack mackerel (*Trachurus
    symmetricus*), 107, 128
Pacific Ocean, 15, 21, 51, 138–9,
    149–52, 238n.49
pain, feeling, 100, 225n.35
Panama, vii, 193
Panamanian isthmus, 51, 142
Papua New Guinea, 94, 140, 188
parasites, 123–5, 127
parrotfish, 113
parthenogenesis, 66, 219n.18
paternity, 63
Perfume (Süskind), 74
Perrin, Jacques, 130

Perryman, Robert, 94
personality, shark 96–9
perspective, changing our, 187–8
Peruvian upwelling, 53, 217n.33
Petrouchka, 51, 53
phenomenal consciousness, 86,
    222n.24
Philippines, 61
phylogeny, vertebrate, 34, *35*, 122
placoderm fish, 43, 62–3
planktivorous fish, 225n.1
planktivorous shark (*Aquilolamna
    milarcae*), 31
plankton, *93*, 94, 133, 137, 206
plastic pollution, 165
Playa del Carmen, Mexico, 88–90,
    89, 95, 166, 171–2
Pliny the Elder, 13, 34, 213n.12
pollution, 164–5, 225n.2
Polynesians, 15, 195
porbeagle shark (*Lamna nasus),*
    156, 238n.4
porite coral colonies (*Porites lutea*),
    30, 71
Port-Camargue, France, *158*
Port Jackson sleeper shark
    (*Heterodontus portusjacksoni*),
    95, 102, 132
Port (killer whale), 131
'post-attack' fishing, 170
potato groupers (*Epinephelus
    tukula*), 109
Prézelin, Louis, 132
primitive beings, 90, 222n.3
purses, 64–5

queen conch (*Strombus gigas*), 60,
    218n.1

rainbow runner (*Elagatis
    bipinnulata*), 129
Raine Island, Coral Sea, 29, 121–2
rays, 31, 34, 40, 73, 81
rebreathers, 175, 206, 244n.21

red scorpion fish (*Scorpaena scrofa*), 64
Red Sea, 194
Reichert, David, 176, 206
remora fish (*Echeneidae* sp.), *124*, 130
reproduction, 61–70
  abandonment at birth, 67, 96
  breeding grounds, 147
  males and, 63, 218n.6
  Pliny on 13
  pregnancy and birth, 60–2, 66–7, 114, 139–40, 171
  sense organs, 76
  species and, 156, 166, 224n.28
Réunion Island, 168–9, 183–6, 195, 243n.11
Richer de Forges, Prof Bertrand, 50
Rio Summit (1992), 199
Risso, Antoine, 149
Rome, ancient, 16
Romulus and Remus, 16
Rondelet, Guillaume, *De Piscibus Marinis*, 15

salinity/density, water, 143, *144–5*, 146, 166, 225n.3, 236n.34–5, 237n.46
salmon shark (*Lamna ditropis*), 107, 234n.6
'salmonette', 163, 164
sandbar shark (*Carcharhinus plumbeus*), 66–7, 120, 173
Sarano, François, 23–5, *24*, 84, *84–5*, *177*
  Lady Mystery and, *208–10*
Sarano, Véronique, 94
Sardine Run, South Africa, 120, 230n.48
sardinella, 110, 112
sardines, 104–6, 117, 120–1, 130, 155, 230n.48
Saving Our Sharks Association, Mexico, 172–3

scalloped hammerhead shark (*Sphyrna lewini*), vi, 58, 95
scallops, 116
Scandinavian mythology, 16
Schluessel, Vera, 96
school shark (*Galeorhinus galeus*), 14, 163
scooter, underwater, 176
scorpionfish (*Scorpaena scrofa*), 64, 154–5, 159
sea lamprey (*Petromyzon marinus*), 124
sea lion (*Neophoca cinerea*), 107, 136
Sea of Cortez, 139
Sea Shepherd, 194
seahorses, 68
Senegal, 160
sense organs, 73–83, *78*, *82*, 99, 220n.6
700 Sharks in the Night (2018 film), 196
sexual maturity, 60, 66–7, 114, 135, 157, 198
Shark Bay, Australia, 142
Shark Education association, 194
Shark Reef Marine Reserve, Viti Levu, Fiji, 165
Shark Safety Centre, Réunion Island, 243n.11
Shark Tale (2004 film), 26, 27
Sharkwater (2008 film), 194
sharpnose sevengill shark (*Heptranchias perlo*), 101
sharptooth lemon shark (*Negaprion acutidens*), 67
shortfin mako shark (*Isurus oxyrinchus*), 107, 117, 157, 161
shrimp, 125, 161
sight, *54*, 79–80, 99
The Silent World (Cousteau), 5, 22–3
silky shark (*Carcharhinus falciformis*), vi, 161

silvertip shark (*Carcharhinus albimarginatus*), 5, 71–2
Singapore, 162
Sioux, 16
skates, 31, *33*, 80
skin, 30, 34, 37, 77, 91, 124, 129
smalltail shark (*Carcharhinus porosus*), 161
smalltooth sand tiger shark (*Odontaspis ferox*), 149–52, *151*, 237n.45
smalltooth sawfish (*Pristis pectinata*), 31
smell, 74–6, 83, 220n.6
smooth hammerhead (*Sphyrna zygaena*), 13, 156
snappers, 109, 113
social network, 94–5
Solmissus (jellyfish), 49
South Africa, 25, 104–5, 132, 183, 230n.48
speciation, 98, 142, 147–9
sperm whale, 15, 17, 92, 95, 98–9, 194
'sphere of intimacy', 84–6
Spielberg, Steven, 25, 48
spines, dorsal, 43, *45*
spinetail mobula ray (*Mobula japanica*), 93–4, 222 n9, 223n.12
spiny dogfish (*Squalus acanthias*), 14
sponges, 13, 157
spotted sea hare (*Aplysia punctata*), 48–9
spurdog shark, *see* spiny dogfish
squalane (shark liver oil), 162–3
Squalus genus, 123
Starboard (killer whale), 131
starfish, 60
Stella (female bull shark), 88–9
Stethacanthus productus (extinct shark-like fish), 44, 46
Stewart, Rob, 194

Stillwell, Lester, 21
stingrays, 53, 72, 91–2, 129
Stort Reef, Sumatra, 91
Strait of Messina, 48, 80
sun ray (*Cyclobatis* sp.), *32*
surfing, 182, 184, 187–8, 244n.17–18
surgeonfish, 109, 113, 118
Surina, Steven, 84, 88–90, 194
Süskind, Patrick, *Perfume*, 74
swallowtail seaperch (*Anthias anthias*), 64
swim bladder, 148
swordfish, 74, 107, 139
Sydney, 182
syngnaths, 68

taste, 76–7
tawny nurse shark (*Nebrius ferruginus*), 8, 67
teeth, 46, 50, *52*
Temaru, Oscar, 195
thermohaline circulation, *144–5*, 146
thinking, understanding each other's, 101–2
thornback skate (*Raja clavata*), 48–9
threatened species, 161, 163–4
thresher sharks, vi, 13, 61, 117, 127
tiger shark (*Galeocerdo cuvier*), 112–13, *114*–15
    attacks on divers, 179, 191
    divers and, 173
    feeding, 29, 122
    Hawaii, 183
    hunted, 160, 167, 184
    Réunion Island, 185
    sexual maturity, 66
Timaru, New Zealand, *42*
Tintin (fictional character), 22
'top-down effect', 115, 119, 228n.30
tope (*Galeorhinus galeus*), 13
tornado, marine, 49, 216n.22

tourism, 172–3, 181–3, 246n.43
transgender fish, 68
Transkei coast, South Africa, 104–5
Les Travailleurs de la mer (Hugo),
	17
trevallies, 126, 171, 192
Trigger (great white), 176, 205
tuna, 16–17, 110, 117, 136, 139,
	155–6, 165
Tunisia, 154, *155*, 156, 157
Tutira, 15
*Twenty Thousand Leagues under the
	Sea* (Verne), 4–5, 17–19
two-spot red snappers (*Lutjanus
	bohar*), 71

Uexküll, Jakob von, 212n.4
Ugly (great white), 176
Umwelt, 2, 73, 96, 212n.4
unicorn fish (*Naso hexacanthus*), 9,
	109, 118

Valdes Peninsula, 142
Valéry, Paul, 121
Vansant, Charles Epting, 20–1
Verne, Jules, *Twenty Thousand
	Leagues under the Sea*, 4–5,
	17–19
vertebrate phylogeny, *35*

Vila Pouca, Catarina, 102
viperfish (*Chauliodus* sp.), 49

Waldie, Peter, 128
Watson and the Shark (Copley), 16
Watson, Paul, 194
Wave (female bull shark), 90
Whale Book (Coenen), 14–15
whale shark (*Rhincodon typus*), 53,
	54, 63, 110, *111*, 112
White Shark Café, Pacific Ocean,
	138–9
white trevally (*Pseudocaranx
	dentex*), 109
whitenose shark (*Nasolamia velox*),
	142
whitetip reef shark (*Triaenodon
	obesus*), 8, 9, *10–11*, 109
Wilcox, Clay, 132
wild animals, 19, 22
wolf, shark and, 15–16
worms, parasitic, 123, 127

yellowtail fusilier, 109

Zarzis, Tunisia, 154, *155*, 157
zebra shark (*Stegostoma fasciatum*),
	65–6
zoos, 198